Lecture Notes in Control and Information Sciences

Edited by M. Thoma and A. Wyner

DFVLR

80

Artificial Intelligence and Man-Machine Systems

Proceedings of an International Seminar Organized by Deutsche Forschungs- und Versuchsanstalt für Luft- und Raumfahrt (DFVLR) Bonn, Germany, May, 1986

Edited by H. Winter

Springer-Verlag
Berlin Heidelberg New York Tokyo

Editor
Heinz Winter
Institut für Flugführung
DFVLR, Flughafen
D-3300 Braunschweig

ISBN 3-540-16658-0 Springer-Verlag Berlin Heidelberg New York Tokyo
ISBN 0-387-16658-0 Springer-Verlag New York Heidelberg Berlin Tokyo

Printed in Germany

Offsetprinting: Color-Druck, G. Baucke, Berlin
Binding: B. Helm, Berlin
2161/3020-54321

P R E F A C E

The German Aerospace Research Establishment - Deutsche Forschungs- und Versuchsanstalt für Luft- und Raumfahrt e. V. (DFVLR) - has initiated a new series of seminars concerning fundamental problems in applied engineering sciences. These seminars are devoted to interdisciplinary topics related to the broad variety of DFVLR activities in the fields of fluid mechanics, flight mechanics, guidance and control, materials and structures, non-nuclear energetics, communication technology, and remote sensing.

The purpose of this series is to bring modern ideas and techniques to the attention of DFVLR staff in order to stimulate internal activities, and to promulgate DFVLR achievements within the international scientific/technical community. To this end, prominent speakers are invited to join in a series of lectures and discussions on topics of mutual interest.

After two seminars

1984 Nonlinear Dynamics in Transcritical Flows, and
1985 Uncertainty and Control

this third seminar deals with "Artificial Intelligence and Man-Machine Systems".

In conventional man-machine systems, the intellectual functions of monitoring, planning, problem solving and decision making are carried out by the operator. Advances in Artificial Intelligence (AI) offer the possibility to support the human operator in these intellectual functions and to improve the performance of the man-machine system.

In this seminar, the state of the art in those domains of AI relating to man-machine systems is reviewed and several selected examples of applications are discussed.

Bonn, May 1986

Prof. Dr. H. J. Jordan
Chairman of the
Board of Directors DFVLR

Contents

Artificial Intelligence in Man-Machine Systems and Automation

Heinz Winter
Institut für Flugführung
DFVLR Braunschweig

1. Introduction

An interesting aspect of the history of mankind is the relationship between man and technology. Technological innovations have had a great influence on human society. A first technological revolution (the agricultural revolution) began with the invention of simple tools, like axes and ploughs. Man learned to use these tools in a manual control loop. He developed particular skills, and subsequently specialization and worksharing began. The effect on society was enormous: The early society of hunters and gatherers transformed into a society of farmers and craftsmen. The development of speech and writing enabled man to codify his skills in the form of rules, and to generalize these rules in the form of knowledge about the tools and their application to the world. The invention of printing made the wide-spread distribution of knowledge possible.

A second technological revolution (the industrial revolution) began with the invention of the steam-engine, which initiated the industrialization of the world. The use of machines gave man the role of a supervisor and monitor of semiautomatic processes. These functions were more based on rules and knowledge than on particular physical skills. An important by-product of this evolution was, that man developed "mental models" of the processes and machines under his control. These models are symbolic representations of his knowledge about the machines and their interaction with the world. More and more skill-based labour was done by machines instead of men. Our industrial society is the product of this development.

A third technological revolution is just going on. It is brought about by the information processing technology and especially by the inventions of artificial intelligence (AI) and knowledge engineering*). These technologies make it possible to implement cognitive capabili-

*)A good introduction to AI is [15]

ties of man, like speech understanding, logic deduction, picture understanding, reasoning about expert knowledge, problem solving and decision making in machines. These inventions will have a dramatic influence on our industrial society, in particular on the sectors of production and administration. It becomes possible to design machines and robots which can do their jobs in an autonomous way without human intervention.It can be foreseen that future space systems will rather be based on silicon and steel intelligence than on flesh and blood. Some people [1,2] see in this development the beginning of the "silicon age", where silicon-based intelligence competes with human intelligence.

Without speculation, however, it can be stated, that the results of artificial intelligence research make the direct implementation of knowledge in computers possible, in a form that this intelligence can interact with machines in a similar way as humans.

Let us now start the consideration of these new technologies and of their role in man-machine systems, with a definition of artificial intelligence, which stems from M. L. Minsky [3]: "Artificial Intelligence is the science of making machines do things that would require intelligence if done by men". This definition refers us back to the human intelligence, for which no generally agreed definition exists. But we all know that human intelligence has to do with the abilities of problem solving and mastering of new demands and situations based on knowledge and reasoning. The pragmatic definition of Minsky describes the most important goal of this research. Another goal is to gain insights into the nature of intelligence and especially of human intelligence.

The development of Artificial Intelligence started in the fifties. In the phase between 1955 and 1970 the basic concepts were developed and small systems were built, which demonstrated the feasibility of artificial intelligence. In the seventies, computer technology provided the means to construct prototype systems, which proved the applicability of the ideas of AI to industrial problems. In the eighties the commercial application of AI has started.

This success was achieved in spite of the fact that many basic problems have not yet been solved satisfactorily. There are several reasons for the enormous interest in AI today:

- Far-reaching R+D programs have been started in different parts of the world to promote AI (DARPA, Fifth Generation and Alvey Programs. Verbundvorhaben [4,5]).

- Hard- and software tools are available which enable the transfer of the ideas to potential users.

- Powerful and cheap computer hardware is available.

In the past thirty years AI research has brought forth important products, e. g. computer languages (LISP, PROLOG [6,7]) for non-numeric (symbolic) data-processing, the technologies of knowledge engineering, expert system shells [8] and hybrid representation languages [9]. The list of successful prototype systems is long. There are

- Expert systems for interpretation and diagnosis (e. g. in medical problems), for systems design (e. g. configuration of computers), for monitoring and control of industrial plants , etc..

- Planning systems, which generate action plans under well defined conditions as well as under uncertainty (for robots or space systems).

- Systems for voice recognition and natural language understanding in limited task domains and well defined contexts.

- Deduction systems for mathematical theorem proving, for automatic programming and program verification, etc..

These are only a few examples of successful applications of AI in different fields. AI is generally considered as a subfield of Computer Science. But there are also many contributions from other scientific disciplines, like Cognitive Psychology, Linguistics and Philosophy.

2. The Structure of Man-Machine Systems

Machines are designed to achieve certain goals, like the mass production of goods, or to enable man to fly from one continent to another, etc.. These goals are layed down in symbolic form (mostly in natural language) as the specifications of the machine. From the specifications a functional description can be developed e. g. in the form of block-diagrams, and finally the physical form of the parts of the machine is defined. So we can distinguish three levels of the description of a machine:

- A **symbolic** representation, which describes the purpose and goals for which the machine is constructed.
- A **functional** representation, describing the functional lay-out of the machine, and finally a
- **physical** representation, which describes the parts, construction materials etc..

One can furthermore distinguish three levels of interaction between an operator and a machine [10]:

- **Skill-based** activities which are carried out almost subconsciously and automatically. A skilled operator has a large repertoire of "automated sensory-motor subroutines" which he can put together to form larger patterns of behavior in order to control the machine. Skill-based reactions are more related to the physical representation of the machine, and the information about the physical state is sensed by the operator in the form of time-space **signals.** The signals, in general, carry no direct symbolic information, but provide a picture of the state and the time-space configuration of the machine and its environment.
- **Rule-based** activities. These interactions are steered by "stored rules" which have been learned during instruction, or which have proven successful in former experiences. In general, these rules cover all situations which have been considered in the design process as routine states of the machine. Rule-based reactions are evoked by **signs.** Signs also do not carry direct symbolic information, like notions or concepts, but they are kind of names for routine states or conditions, which can be handled by routine reactions of the operator.

In case of unexpected events, like system failure, no rules may exist to cope with the situation. These cases must be mastered by the operator with

- **knowledge-based** activities. He must now rely on his knowledge about the goal to be achieved by the machine and about the functional lay-out and the physical form of the machine. On this high conceptual level, he must be able to develop a plan to solve the problem and subsequently implement this plan. Knowledge-based reactions require symbolic information which is perceived as **symbols**. Symbols refer to concepts like goals, functional properties etc., and are the basis for reasoning processes. The difference between signs and symbols can be elucidated by the statement that signs are part of the physical world, whereas symbols are part of the human world of meaning.

This (simplified) scheme of man-machine interactions is illustrated in Figure 1. It is important to note, that the destinctions between skill-, rule- and knowledge-based man-machine interactions and also between signals, signs and symbols are not absolute. For example, knowledge-based activities - if carried out frequently- can become routine and assume the form of a new rule; similarly rule-based activities can become skill-based actions.

We have seen that knowledge about the machine and its environment, the "world", exists in physical, functional and symbolic forms. It is still a subject of research to-day to find out, if human operators use similar representations in their mental models. It seems that humans have the ability to switch beween models of different structure and of different abstraction level as the situation requires.

3. Automation in Man-Machine Systems

In this paper the term "automation" is used to describe the effort to transfer human activities (physical or mental) to machines, with various levels of remaining human participation. Today most of the physical human work is transferred to machines. The role of man has shifted from a system controller to a system monitor and problem solver. Modern machines - aircraft can be considered as an example - are highly automated and can achieve their goals in almost the whole range of the conditions foreseen by the designer without human intervention. The human operator supports the machine

- by maintaining the conditions which are required for normal and satisfactory operation, and

- by intervening in such situations which could not be foreseen by the designer, or in the case of malfunctions or accidents.

This means that the human activity has shifted from skill- and rule-based to knowledge-based functions, such as reasoning, planning and problem solving. This situation is illustrated in the Figure 2.

This degree of automation was made possible mainly by optimal control theory. With the methods and tools of this discipline, optimal controllers can be designed which take over skill- and also some rule-based human activities. Optimal control theory fails, when no algorithmic rule is known which solves the problem and when "heuristic" rules have to be applied. In this case knowledge-based methods can provide the solution.

Artificial intelligence technology offers the potential to transfer the higher intellectual functions (rule- and knowledge-based) from humans to computers. In the long term this can lead to completely autonomous machines, which have the ability to cope with unforeseen events - such as failures of subsystems - and which can repair themselves or if repair is not possible continue to function on a lower performance level.

In the remaining part of this paper we consider machines, which have the capability of autonomous operation, but which are not expected to work only in a completely autonomous mode. Such systems - again aircraft are a good example, but also air traffic control systems and certain space vehicles - will work together with humans, at different levels of autonomy. This leads to particular design features of the machines, which allow the operator to intervene on all the three levels of the man- machine interaction discussed above.

One key to a successful realization of these principles is to find symbolic representations, which enable the operator to understand what is going on in the machine. In normal and routine situations he reacts in a skill- and/or rule-oriented way. He switches to a knowledge-oriented behavior if new and unforeseen events occur. For his activities he uses physical, functional and symbolic mental models of the machine and the world. For man-machine systems with the capability of both autonomous and manual operation it is important to find representations in the computer which are similar to those used by the operator. This design principle can be called anthropomorphism. A machine, with anthropomorphic design can be considered as a "silicon" double of the operator (examples are the Pilot's Associate [11] or the Astronaut's Apprentice [12]).

Before we go to a deeper consideration of the automation process, let us have a look on problems observed in past automation processes. Again the situation in an aircraft will be taken as an example. About 50 years of experience with aircraft accidents and incidents are at hand. During this time, aircraft have changed from relatively simple manually flown machines to the highly automated and complex systems of today - like the A310, A320, B757 and B767. Research and studies [13] indicate, that with increasing degree of automation, the percentage of accidents caused by "man" has increased, whereas the percentage of "machine"-caused accidents has decreased. Also the nature of the human errors responsible for accidents has changed. In non-automated aircraft, pilot errors were more oriented towards misreading of displayed information and manipulating the wrong controls; in automated aircraft data-entry errors, misinterpretation of the reactions of automatic systems in non-routine situations and misprogramming of automatic systems seem to be important sources for man-caused accidents. This indicates two conflicting tendencies:

- More automation promises to reduce the likelihood of human errors.
- More automation creates new possibilities for human errors.

The way out of this problem is a better design of the man-machine system - both of the system structure and of the interfaces between man and machine. Conventional automated systems, like those in the present aircraft are to a large extent based on optimal control theory. It is a great hope, that automation architecture based on artificial intelligence, on the principle of anthropomorphism and on a proper distribution of authority between man and machine can increase the safety of the total system and at the same time improve its performance. Several important rules have to be observed in the design of these man-machine systems:

- One must keep the man in an appropriate level of alertness when he is monitoring the system.
- When the system works in the automatic mode, man must "stay in the loop" in order to minimize reaction time for take-over in case of system failure.
- Man must be able to intervene at all three levels of interaction (skill-, rule- and knowledge-based) in order to serve as a back-up in case of a failure of the automated system.
- The display and interface systems must be laid out to ease communication between man and machine on the three levels of interaction.

A paradigmatic description of an automated man-machine system, which is capable of operating with various levels of human participation, is shown in Figure 3. The interface between the man and the machine allows both automatic and manual operation. The optimal controller has been supplemented by artificial intelligence, which duplicates the human capabilities of knowledge- and rule-based reasoning. An "intelligent interface" handles the communication between the man and the machine intelligence. The interface must have intelligence, because it must be able to understand the intentions of the operator, e. g. the selected level of automation.

An example of 5 different levels of authority distribution between the operator and the machine intelligence is this one (see also [14], where ten automation levels are described):

Level 1: Human does the planning, makes the decision between suitable plans and implements the decision (manual mode).

Level 2: Machine does the planning, offers a restricted set of suitable plans to the human, and human decides on one of these and implements it.

Level 3: Machine does the planning, suggests a plan to the human, and implements it, if the human approves.

Level 4: Machine does the planning, decides on a plan, implements it and informs the human after the fact.

Level 5: Machine does the planning, makes the decision between alternative plans and implements the decision (autonomous mode).

Present man-machine systems typically operate on Level 1, with little or no machine intelligence. A man-machine system with implemented artificial intelligence (like the one in Figure 3) should be able to operate in all the five levels of authority distribution.There are several possibilities for the adjustment of the level of automation. It can be done by the human operator, or by a "meta-controller" (Figure 4). The meta-controller must have knowledge about the human operator, the machine and the surrounding world. Based on this knowledge, he decides on an optimal workshare between the human operator and the machine and assigns their tasks. The meta-controller can be a human (like a ground-station in space activities) or "meta-intelligence" implemented in a computer.

4. The Role of Artificial Intelligence

In the preceding chapters we have seen that artificial intelligence and knowledge engineering provide the potential to transfer the higher cognitive functions of man to the machine. Examples of these cognitive functions are:

- Monitoring the state of the machine and the surrounding world.

- Goal detection. Goals may arise from changes of the operational conditions of the machine, from unforeseen events and from malfunctions.

- Plan development. A set of suitable plans is developed to react on the observed events and to reach the detected goals.

- Plan projection. The effect of alternative plans on the system must be evaluated in a kind of "simulation".

- Decision on the optimal plan and implementation of this plan.

These activities, if implemented in a machine (a computer) require certain basic capabilities, which are studied in artificial intelligence research:

- The representation of knowledge about the physical form, the functional structure and the purpose and goals of the machine and about the surrounding world, and the ability to reason about this knowledge. A good overview on knowledge representation schemes and their use is given in [16]. In their paper in this volume, Mylopoulos et al. describe the knowledge representation language Taxis and its application in the software development process.

- "Expert" knowledge plays an important role, which is captured in the form of rules in so-called "expert systems" - another discipline of artificial intelligence. In his paper in this volume, Radig describes the basic concepts and architecture of expert systems and illustrates their application.

- Goals, symbols and knowledge are often expressed in natural language; so the ability to understand natural language is a prerequisite for the duplication of human abilities in the machine. This capability has another advantage: The communication between human and machine can be made in natural language. Wahlster's paper in this volume gives an introduction to natural language analysis and understanding and presents the results of the evaluation of three commercial natural language systems.

Not very much is understood so far of the processes in the human brain, when the above mentioned cognitive functions take place. Cognitive Science, another scientific discipline associated with Artificial Intelligence, studies the relationship between information processing in humans and in computers. Theories, methodologies and results from Cognitive Science, applications to human-computer systems and intelligent user support systems are described in the paper of Fischer in this volume.

A problem arises, when the knowledge on which a decision or a plan has to be based is vague, incomplete or uncertain. The human is able to plan and act under such conditions. Artificial intelligence research has developed models and theories to implement this capability into computers. [15] gives a good introduction in the certainty factors approach to this problem. Zadeh's paper in this volume presents an approach to reasoning under uncertainty in the framework of fuzzy logic and its application in expert systems.

Another difficulty comes from the fact that the knowledge-based processes are generally non-deterministic in nature. In classical optimal control theory, if all the inputs to a system and its internal structure (including the initial state) are known, the future state development of the system is determined. This is not necessarily the case in knowledge based systems. Consider for example non-exhaustive search processes in a knowledge base. If the search is guided by heuristic information - which may even depend on the actual state of the knowledge base - the result of this search process is not predetermined at all! This fact poses problems in safety critical systems, where the safety and reliability has to be demonstrated to legal authorities in verification and validation processes. This situation is similar to the verification and validation of human performance.

5. Examples of Application

The implementation of artificial intelligence in technical systems progresses in a breath-taking speed. Expert-systems especially find wide-spread applications. In the field of man-machine systems which we consider in this seminar many projects have been initiated in the past years. The Pilot's Associate [11] and the Astronaut's Apprentice [12] have already been mentioned as examples of "silicon doubles" of human experts in aeronautics and astronautics. Meystel's paper in this volume discusses the application of AI techniques to the control of mobile robots. In the paper of Cross et al. the use of AI for tactical mission planning in the cockpit of a military aircraft is described.

Several projects have also been started within the DFVLR. The introduction of artificial intelligence into the cockpit of transport aircraft is studied in the project "Flight Operations System (FOPS)".

FOPS can be considered as an extension of the Flight-Mangement-System (FMS) of present transport aircraft (Figure 5). It is designed as an intelligent interface system between the pilot, the aircraft and air-traffic control (Figure 6). The capability of speech recognition and understanding with limited vocabulary, syntax and semantics shall make an easy communication between pilot, aircraft and Air-Traffic Control (ATC) possible. FOPS converts the pilot's voice inputs into machine commands which activate the corresponding on-board systems. One of these subsystems is the digital data-link between the aircraft and the ground-based ATC. In a later phase, FOPS will have a reasoning capability and a knowledge base about the aircraft and the surrounding traffic, which enables it to check the spoken messages against the current situation of the aircraft and the surrounding world, in order to detect misunderstandings and errors. The FOPS will be part of the Experimental Cockpit (Figure 7) and will be flight-tested in the ATTAS (Advanced Technologies Testing Aircraft System) research aircraft of DFVLR.

Another project is the development of an intelligent self-calibration and fault-identification system for avionics. The ATTAS airplane carries multiple inertial, radio and air-data navigation systems. The outputs of these systems are fed into a Kalman filter to obtain optimal estimates of the aircraft states, its acceleration, position and velocity. The sensor outputs together with the optimal estimates are introduced into a knowledge based reasoning system, which contains expert knowledge about the senors' failure characteristics. This expert system identifies malfunctions of components and sensors, eliminates false sensor outputs and sends a message to the flight-engineer.

An important area of application of artificial intelligence in man-machine systems is air traffic control. To-day, air traffic is handled by the controllers in a completely manual mode. Together with the German air traffic control authority BFS (Bundesanstalt für Flugsicherung), DFVLR initiated the COMPAS (Computer Oriented Metering Planning and Advisory System) project several years ago. The aim is to make a first step in the introduction of intelligent computer assistance to the air traffic controllers. The basic ideas of this project are reported in the paper of Völckers in this volume. Based on an intelligent search process, an automatic planning capability proposes a strategy to control the flow of incoming aircraft at Frankfurt Airport. Much effort in this project has been spent on the optimization of the interface to the controller, which, in the project, is based on present technology.

Manned space missions are another field of application. For the German Spacelab mission D2 a robotics experiment is foreseen, which shall use several intelligent components:

- A teach-in capability, through which the astronaut can "teach" the robot to perform certain tasks.

- A health and fault management system, which uses an expert system for diagnosis of error or failure situations and which proposes recovery actions to be taken by the astronaut.

6. Conclusion

It has been shown that artificial intelligence offers the potential to transfer the knowledge-based functions of a human operator to the machine. In systems - such as aircraft, air traffic control and manned space stations - where the human ability to back-up all control functions should be retained, it is important to design the man-machine system in a way to allow operation on different levels of automation (from manual to autonomous). This keeps the human "in the loop" and enables him to "take over" immediately in the case of emergency or system failure. An interesting aspect of the system lay-out is that it allows different locations of the meta-controller. This is the intelligence above the man-machine system, which decides on the goals, on the authority distribution between the man and the machine and has the most complete knowledge of the world in which the man-machine system operates. The meta-control can be assigned to a human or a computer, it can be collocated with the machine or remote (e. g. the ground station in space experiments).

References

[1] Van Cott, H. P.
From Control Systems to knowledge Systems
Human Factors, February 1984, 26(1), p. 115

[2] Sir Clive Sinclair on the Third Industrial Revolution.
Future Generations Computer Systems (FGCS)
Vol. 1, No. 2, p. 119, 1984

[3] Minsky, M. L.
Artificial Intelligence.
Freeman, London, 1966

[4] Feigenbaum, E. A. / McCorduck, P.
Die Fünfte Computer-Generation
Birkhäuser, Stuttgart, 1984

[5] Bekanntmachung über die Förderung von Forschungs- und Entwicklungsvorhaben auf dem Gebiet der Informationsverarbeitung
Bundesanzeiger, 21.3.1984

[6] Winston, P. H. / Horn, B. K. P.
LISP
Addison-Wesley, Reading (MA), 1981

[7] Clocksin, W. F. / Mellish, C. S.
Programming in Prolog
Springer, Berlin, 1981

[8] Hayes-Roth, F. / Waterman, D. A. / Lenat, D. B.
Building Expert Systems
Addison-Wesley, Reading (MA), 1983

[9] di Primio, F. / Bungers, D. / Christaller, F.
BABYLON als Werkzeug zum Aufbau von Expertensystemen.
in "Wissensbasierte Systeme", Informatik Fachberichte 112,
Springer Berlin, 1985, p. 70.

[10] Rasmussen, J.
Skills, Rules, and Knowledge; Signals, Signs, and Symbols, and Other Distinctions in Human Performance Models
IEEE Transactions on Systems, Man, and Cybernetics, VOL. SMC-13, No. 3, May/June 1983

[11] Stein, K. J.
DARPA Stressing Development of Pilot's Associate System.
Aviation Week + Space Technology, April 22, 1985, p. 69.

[12] Cliff, R. A.
Program Plan for the ASTRONAUT's APPRENTICE.
An Evolutionary Program for the Purpose of Producing Highly Capable Human-Machine Systems in Space
Unpublished Contractor Report, 1985

[13] Chambers, A. B. / Nagel, D. C.
Pilots of the Future: Human or Computer?
Communications of the ACM, Vol. 28, No. 11,
Nov. 1985, p. 1187.

[14] Klos, L. C. / Edwards, J. A. / Davis, J. A.
Artificial Intelligence - An Implementation Approach for Advanced Avionics.
in "AIAA Computers in Aerospace IV Conference" October 24-26, 1983, Hartford (Conn.), p. 300.

[15] Winston, P. H.
Artificial Intelligence
Addison-Wesley, Reading (MA), 1984

[16] Mylopoulos, J. / Levesque, H.
An Overview of Knowledge Representation.
in Brodie/Mylopoulos/Schmidt (eds.) "On Conceptual Modelling: Perspectives from Artificial Intelligence, Databases and Programming Languages", Springer-Verlag, 1983

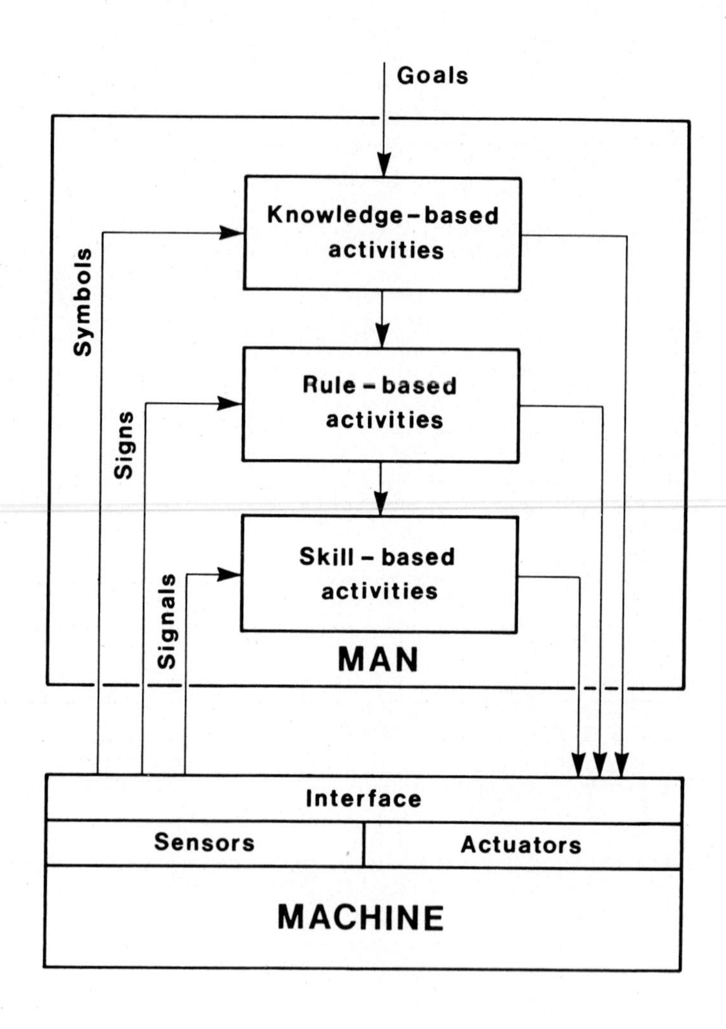

FIGURE 1: SIMPLIFIED DESCRIPTION OF A CONVENTIONAL MAN-MACHINE SYSTEM.

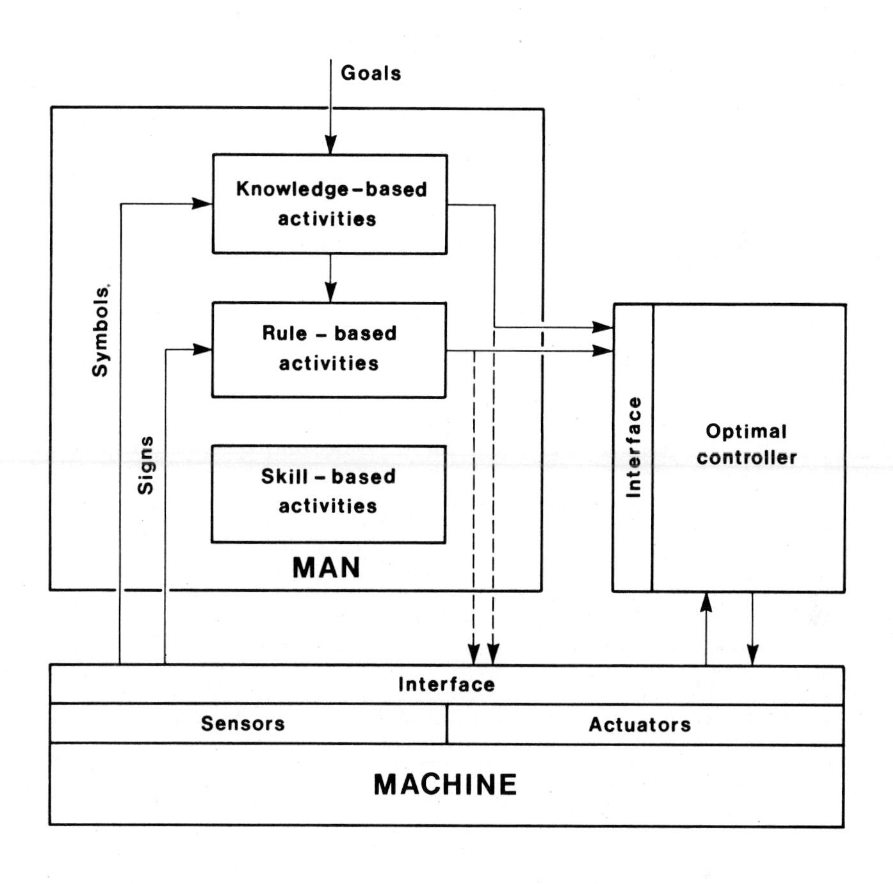

FIGURE 2: PRESENT LAY-OUT OF AUTOMATED MAN-MACHINE SYSTEMS.

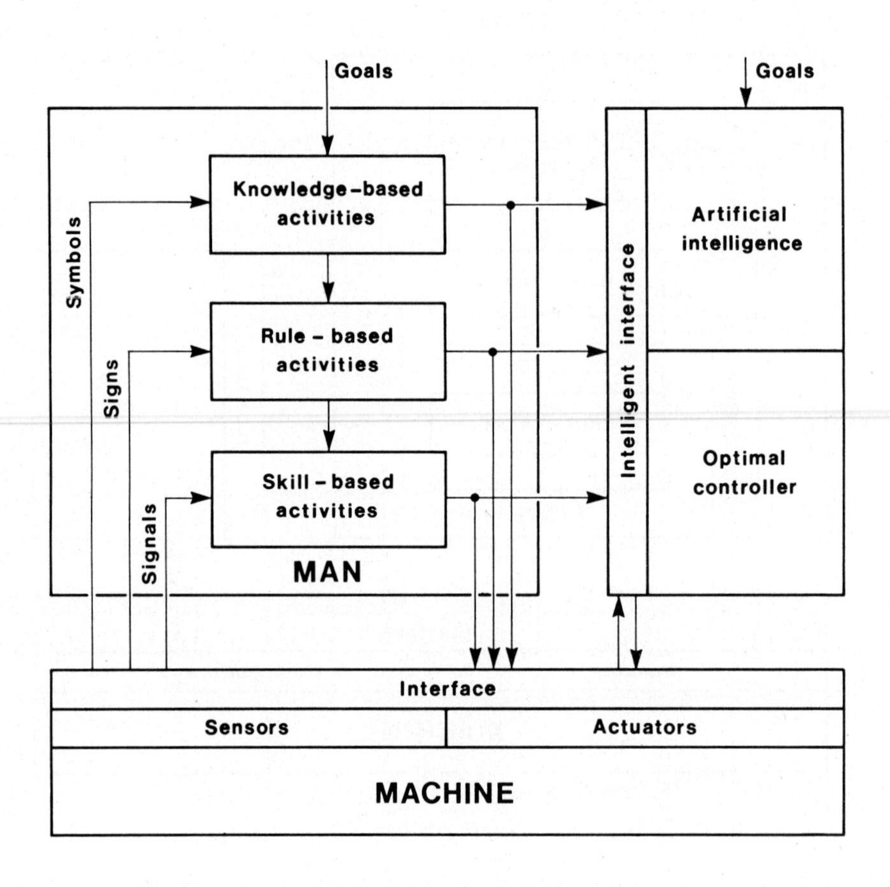

FIGURE 3: PARADIGMATIC DESCRIPTION OF AN AUTOMATED MAN-MACHINE SYSTEM WITH ARTIFICIAL INTELLIGENCE.

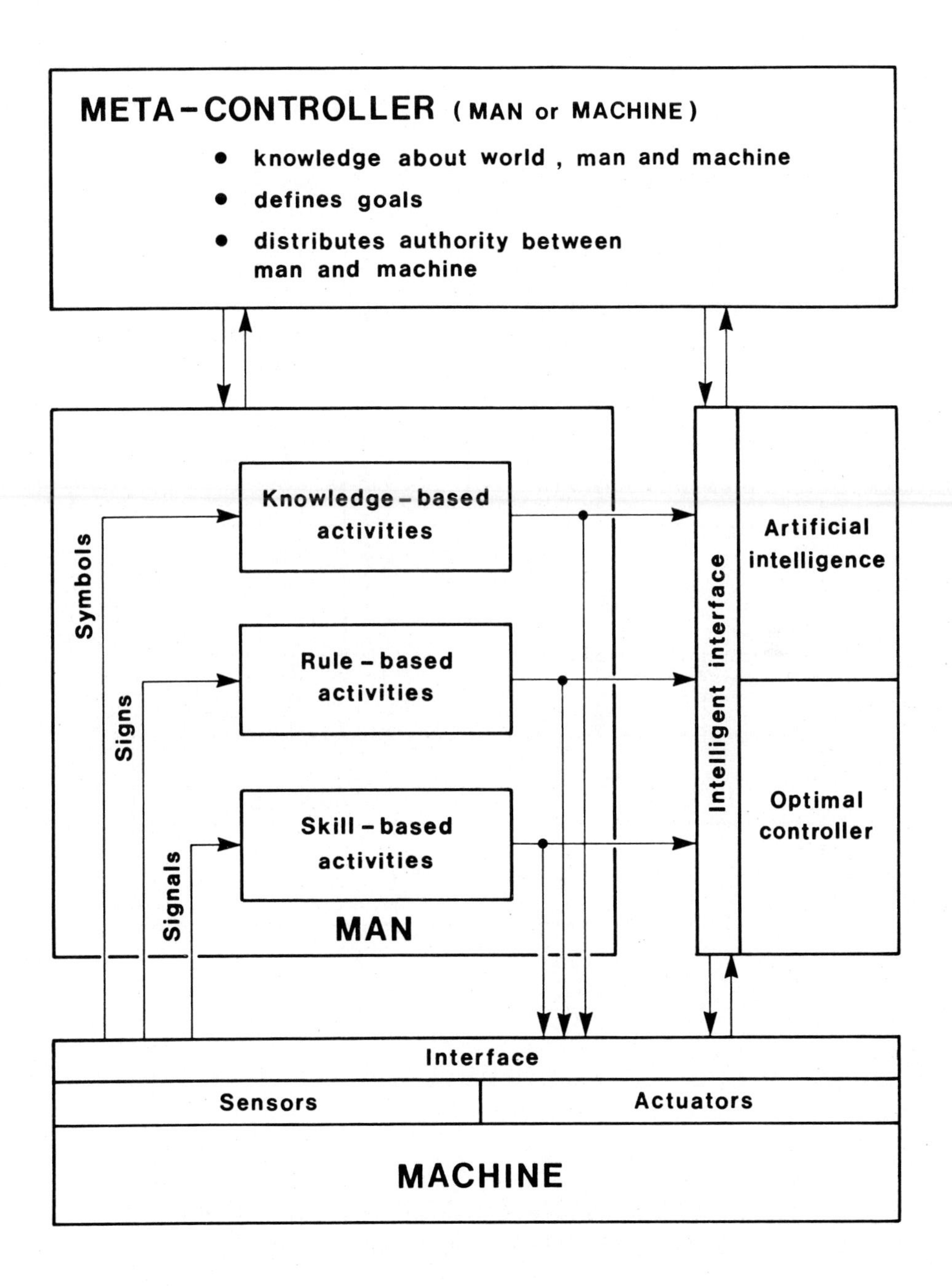

FIGURE 4: SYSTEM WITH VARIABLE DISTRIBUTION OF AUTHORITY BETWEEN MAN AND MACHINE.

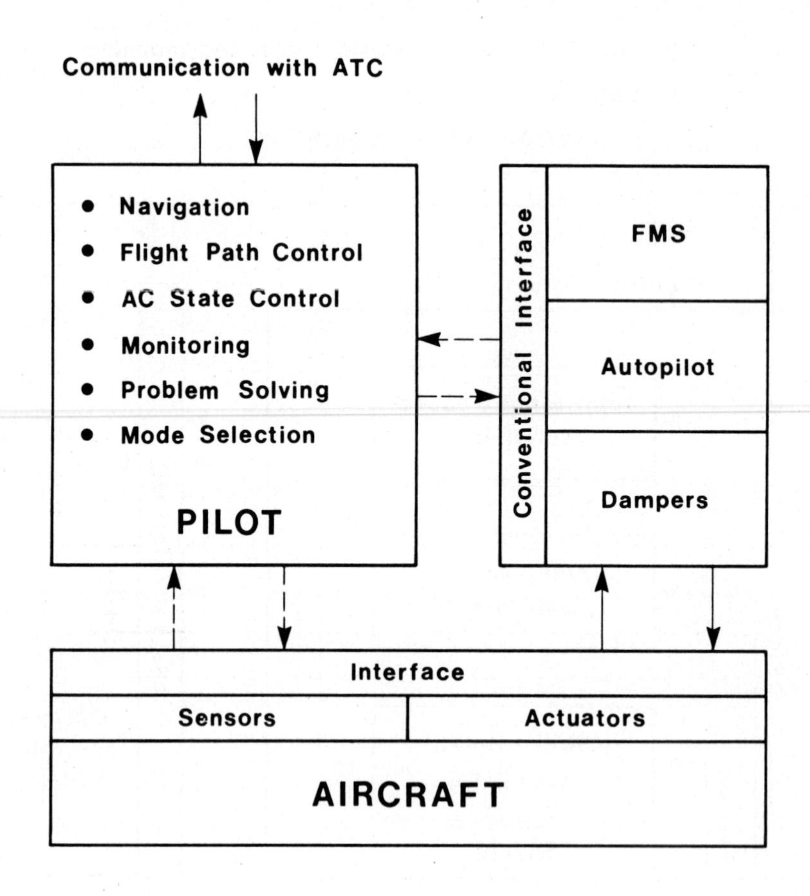

FIGURE 5: PRESENT WORKSHARE BETWEEN PILOT AND AUTOMATIC SYSTEMS IN TRANSPORT AIRCRAFT.

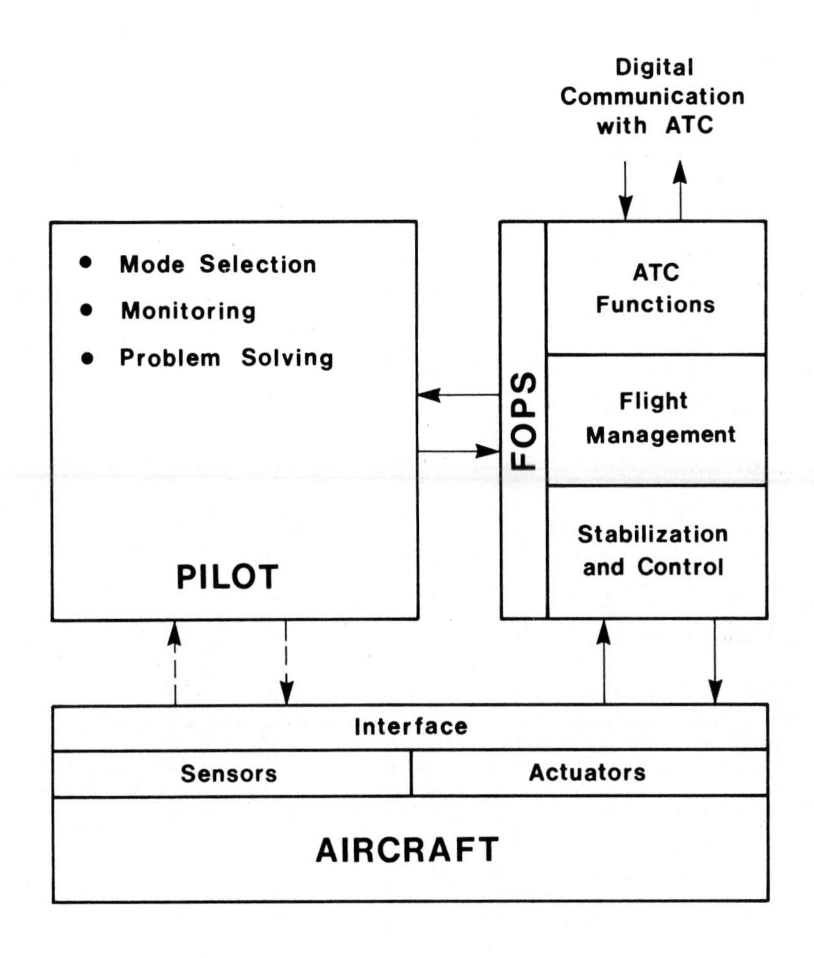

FIGURE 6: FLIGHT OPERATIONS SYSTEM (FOPS) IN A TRANSPORT AIRCRAFT.

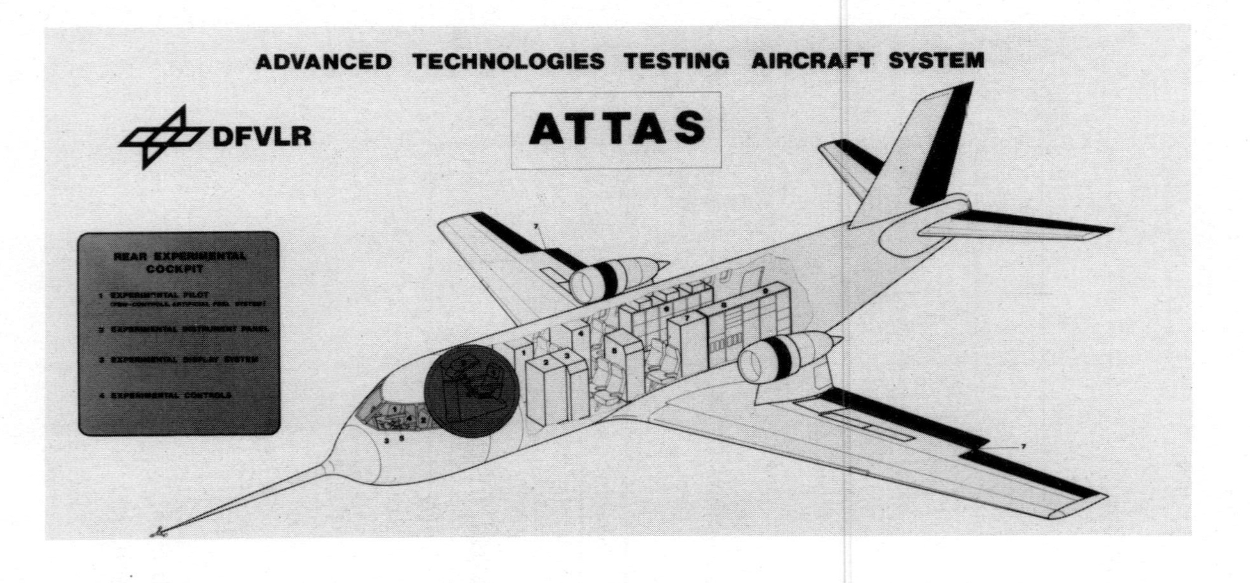

AI RESEARCH PROJECTS

- EXPERIMENTAL COCKPIT
- FLIGHT OPERATIONS SYSTEM (FOPS)
- AVIONICS SELF - CALIBRATION AND FAULT - IDENTIFICATION SYSTEM

FIGURE 7: ATTAS EXPERIMENTAL COCKPIT AS TESTBED FOR INTELLIGENT AVIONICS.

Knowledge Representation in the Software Development Process: A Case Study[1]

John Mylopoulos[2]
Alex Borgida[3]
Sol Greenspan[4]
Carlo Meghini[5]
Brian Nixon

Department of Computer Science
University of Toronto
Toronto, Ontario, Canada M5S 1A4

ABSTRACT

This paper presents an overview of the premises, research results, prototype implementations and prospectus of the Taxis Project at the University of Toronto.[6] The project addresses several problems of Software Engineering within a unified representational framework. The framework offers a coherent set of basic principles for *modelling* and *abstraction* -- drawn from work in *knowledge representation* in the field of Artificial Intelligence (AI) -- and applies them to several aspects of system design and description.

1. Introduction

1.1. Toward a Unified Approach to Software Engineering

Software engineering traditionally involves a number of system descriptions, or *specifications*, which serve different purposes. The types of specifications can be thought of as addressing issues of description at different *levels*:

System requirements specification -- a description of the overall function that needs to be performed by some system.

System design specification -- a high-level description of the system architecture, i.e. the structure of the system in terms of interconnected modules.

Program specification -- the algorithms and data structures, described in a manner independent of what programming language(s) will be used.

Program -- a set of instructions and data, encoded in a programming language, to implement the program specification.

From requirements through programming, these levels range from implementation-independent specifications to implementation-oriented ones. Requirements are implementation-independent in that they deal with concepts of one or more application domains and describe the purpose to be served by a system in a real-life environment. The successive categories move increasingly closer to descriptions of a functioning computer system, expressed in terms of concepts of computation.

---

[1] An earlier version of this paper was presented at the 1983 National Conference of the Canadian Information Processing Society in Ottawa.

[2] Senior fellow, Canadian Institute for Advanced Research.

[3] Current address: Department of Computer Science, Rutgers University, New Brunswick, NJ 08903, U.S.A.

[4] Current address: Schlumberger-Doll Research, P.O. Box 307, Ridgefield, CT 06877, U.S.A.

[5] Current address: Istituto di Elaborazione della Informazione, CNR, Via Santa Maria 46, I-56100 Pisa, Italy.

[6] As of December 1985.

A *methodology for software engineering* has traditionally been expected to provide at least three things for each level of specification:

languages --convenient notations for expressing the relevant information.

techniques -- procedures for constructing, manipulating, and validating specifications.

tools -- automated aids designed to support the above.

Advances in program design and programming include high-level language facilities, techniques such as structured programming, stepwise refinement, and program analyzers, and programming support facilities. The same kinds of technology are needed for requirements and design, if the overall task of software engineering is to be done efficiently and predictably.

Technological advances in programming have resulted from the ability to *understand* and *formalize* program concepts and the programming process. However, requirements and design are not as well understood, and due to their world-oriented nature, formalization is very difficult. Better languages, techniques, and tools need to be developed. By a unified approach to software engineering, we mean one that combines specifications from all levels within one methodological framework.

At each level of specification, we need to describe at least the following aspects of systems:

Information -- the information that is stored in, or can be extracted from, the system.

Behaviour -- the prescribed ways that information can be changed over time.

User interface -- the interaction between the system and users.

Even a cursory glance at the state-of-the-art shows that a number of descriptive approaches have been proposed and/or used, addressing different specification levels and different system aspects. For requirements and design, some of the prominent languages and tools are PSL/PSA [Teichroew77], used for information system description, RSL and its associated tools [Alford77], which are used for real-time systems, and SADT (a trademark of Softech, Inc.) [Ross77a, Ross77b], which has been widely used, mainly for requirements definition. At lower levels of specification are some system design and program specification techniques, e.g. [Jackson83].

In response to the critical need to manage large quantities of information, many organizations have turned to description techniques such as the Entity-Relationship approaches based on [Chen76], and, in the area of Database Management, data models are popular devices for descriptions at different levels of database design and implementation [Tsichritzis81]. There has also been a lot of work in the area of software specifications (e.g. [SRS79]) which, in contrast to the data models, have concentrated on behaviour rather than information modelling.

To summarize, current approaches address particular aspects at particular levels, but the crucial problem remains: *How can an appropriate set of modelling facilities be combined into a unified framework?*

1.2. The Role of Knowledge Representation in Software Development

The development of useful software demands the integration of large amounts of multifaceted knowledge that needs to be extracted from many different sources and reshaped into a functional computer program. For example, a student registration system needs to "know" that students have an associated department attribute and courses an instructor, also that students enrol in courses, if certain rules are satisfied. The knowledge relevant to a particular software development task may best be captured in terms of a specialized language. Indeed, large classes of such tasks involving scientific computing have been handled relatively successfully in terms of conventional programming languages starting with FORTRAN and ALGOL 60. But for many other tasks, conventional and not-so-conventional languages offer an inadequate basis for capturing and representing relevant knowledge thus exacerbating the so-called software crisis.

Over the past decade Artificial Intelligence (AI) has made some progress towards the development of concepts, linguistic tools and techniques for the representation of knowledge. **It is our basic thesis that software development tools, methodologies and environments can benefit by adopting the fruits of this research.**

A fundamental aspect of Knowledge Representation research has been the study of the nature of *knowledge, belief and conjecture,* their formal properties and their use in reasoning, planning and interpretation, e.g., [Levesque84], [Delgrande85], [Konolige85]. Another aspect of this research concerns the study of particular kinds of knowledge, such as spatial, temporal or causal knowledge, e.g., [Allen81], [Shibahara85]. Turning from foundational concerns to engineering ones and noting that useful knowledge bases will in general be large, Knowledge Representation research has examined structuring mechanisms for knowledge bases that suggest the use of ***taxonomies*** (or ***is-a hierarchies***) of concepts, ***aggregation hierarchies, contexts*** and the like (see [Mylopoulos80a] for a survey). Mathematical Logic has served as yardstick and inspiration for much of the research on Knowledge Representation, both in terms of its linguistic constructs and its methodology for defining and studying a formal calculus. Semantic nets have also provided useful data structures for representing and structuring knowledge bases while procedural representations employing a variety of activation mechanisms and control structures have been developed and used in the construction of useful, i.e. efficient, "knowledge based" systems. It is fair to conclude this discussion with an admission that we have only started to scratch the surface in our understanding of knowledge and how to capture and represent it computationally.

In examining software specifications, we propose to distinguish between ones that serve as descriptions of some "world" or "slice of reality", and ones that serve as full descriptions of the intended behaviour of a mechanism [Hehner85]. The former provide a ***model*** by capturing knowledge about the real world, and are generally incomplete, imprecise and possibly inaccurate; the latter are descriptions of systems and focus on the ***function*** or behaviour of the systems. The importance of the distinction between specifying models and functions during the software development process is thoroughly discussed in [Jackson83]. By-and-large, however, this distinction has either been dismissed or ignored by software engineers. We believe that the distinction is all-important and that the software development process should begin with world modelling.

The application of Knowledge Representation to world modelling is obvious and needs little additional elaboration. Its application to function specification, however, is slightly more problematic and requires some discussion. One might argue that system function specification should be independent of the environment within which the system under development will eventually function and should simply provide an account of the system's behaviour. [Hehner85] provides one of the more recent, and elegant, examples of this approach. Consider, however, an interactive information system which maintains information about, say, students at a University. A formal specification of such a system which merely describes its behaviour but doesn't try to provide an account of what the information handled by the system ***means*** in the first place, i.e., how it relates to reality, seems incomplete. It tells us how symbols will eventually be pushed around inside a machine. It doesn't give us any guidelines on how to interpret those symbols. This observation suggests that, at least for information systems, a functional specification should come with an account of how mechanism behaviour corresponds to the world being modelled. For functional specifications of this sort we need linguistic tools which on one hand allow the description of system components, their states and I/O behaviour, on the other come with a ***rich semantic theory*** that allows one to relate system states and function to the world being modelled. So-called semantic data models (attempt to) do just that.[7]

1.3. The Taxis Project

Within the Taxis Project at the University of Toronto, modelling for both requirements and information system design is addressed within the same descriptive framework. The Taxis framework offers a small set of basic representational principles which are applied uniformly to all aspects of modelling. We believe that such a framework is essential to the development of an effective software engineering methodology. The basic concepts are discussed in Section 2.

The focus of the Taxis Project has been the Taxis language, which was developed to provide a framework for information system design. The language offers a notation that supports its underlying principles, providing a natural way of encoding knowledge about the application domain. Taxis integrates, within a framework of classes, properties, and inheritance, the notions of data classes, which are similar to relations [Codd70]; standard Algol/Pascal control structures; and procedural exception-handling as in [Wasserman77]. Taxis is intermediate between a nonprocedural conceptual requirements specification language and an ordinary application language plus database management system as used today.

The linguistic facilities offered by Taxis allow one to design an information system by defining and structuring a large collection of classes. Some classes represent static information, e.g., the students known to the system at a particular moment, others dynamic, such as the enrolment procedure or the (long-term)

[7] See [Brodie84] and [Borgida86] for surveys.

event of studying at a University. One of the aims of the Taxis project is to develop a compiler that translates Taxis programs to a high-level programming language with relational database facilities, such as Pascal/R [Schmidt80].

The facilities of the language can be classified into three categories depending on the kind of knowledge they are meant to represent:

> The *semantic category* includes features that allow one to describe in a model the entities and relationships constituting the application domain, thus defining a database and its associated transactions. The user can specify integrity constraints, exceptions, and exception-handling within the language.
>
> The *pragmatic category*, including mostly Taxis *scripts*, addresses the needs of modelling the interaction between users and the system. Scripts provide a graphical language based on Petri nets, combined with communication primitives, for modelling the flow of events in a user's world and the points of interaction with the system.
>
> The *linguistic category* of Taxis features includes facilities for designing languages (e.g. query languages and natural language front ends) that allow communication between the system and the end user.

We discussed earlier our conviction that software development should start with a world model which provides a formal account of the world within which the software under development will eventually function. For this, the Taxis project has been working on the development of a second language:

> A *requirements model* is a description of some portion of the world that encompasses potential information systems and is used to communicate and analyze the problem situation; it also provides a starting point for information system design using Taxis.

A model at this level corresponds to "Corporate Requirements" as defined in [Lum79].

The requirements modelling language RML [Greenspan84] exemplifies the kinds of linguistic tools one can develop using knowledge representation techniques at this level. RML treats a requirements specification as a **history** of the entities and activities constituting the world being modelled.

We will discuss the Taxis language components and RML in Section 3.

2. Foundations

The Taxis representational framework consists of two main ingredients: an object-oriented framework and a set of abstraction principles. In this respect Taxis is similar to other object-oriented frameworks, such as those of Simula [Dahl72] and Smalltalk [Goldberg81]. In such a framework, a model consists of a set of interrelated *objects*. Each object in the model is intended to stand for some entity or concept in the world or system being modelled. The creation, modification, and manipulation of objects is taken to represent the behaviour of their counterparts (referents) in the world. A big advantage of object-oriented frameworks is their ability to project a natural correspondence between the model and the world. An important difference between the Taxis and Smalltalk frameworks is that Smalltalk treats **message passing** as the fundamental feature of objects while Taxis emphasizes object organization along different abstraction dimensions over message passing.

The most prominent collection of object-oriented frameworks in AI is based on *semantic networks* [Quillian68, Findler79], which are popular schemes for representing knowledge to be used by knowledge-based systems such as "expert systems". We believe these schemes have a lot to offer in the area of Software Engineering for modelling and specifications. The Taxis framework itself derives from Procedural Semantic Networks [Levesque79] and has the benefit of borrowing insights and results from this and related work in AI.

The second basic ingredient of the Taxis framework is a set of *abstraction mechanisms*, which are important in a modelling language for structuring and organizing large descriptions.

An abstraction mechanism is a conceptual or linguistic mechanism that allows certain information to be highlighted while suppressing other information. In software engineering, abstraction is usually equated with the suppression of design decisions or implementation detail along the dimension we have been referring to above as "levels of specification". However, within a given level, the Taxis framework offers a set of complementary abstraction facilities based on the notions of *aggregation, classification, and generalization* [Smith77].

If we define a *property* to be a directed relationship between two objects, aggregation allows one to view an object as a composite of the objects to which it is related by properties. For example, a person has a name, an address, and so on. The "abstraction" here is that one may talk about an object while choosing to ignore its components for the moment.

Along an orthogonal dimension, the classification abstraction allows individuals to be grouped into classes (and classes into *metaclasses*) that share common properties. A class represents a generic concept, such as "person" or "employee", and it also serves as a template for the members of the class, which are called its *instances*. The class is defined by *structural* (also called *definitional*) *properties*, which express such information as "persons have a name, an age, address" and so on. Instances have *factual properties* which, for example, attribute specific names, ages, etc. to individual persons. The "abstraction" here is that one may describe a class without referring to its instances.

Generalization (and its converse, *specialization*) allow the common properties of several classes to be abstracted into the definition of a single, more general, class. For example, the class of persons can be represented as a generalization of the classes representing males, females, managers, engineers, female engineers, and so on.

Accordingly, classes are organized into a hierarchy with general classes located above their specializations. If one class is defined to be a specialization, or *subclass* of another (a *superclass*), every instance of the first is considered to be an instance of the second, e.g., every instance of the class of employees is an instance of the class of persons. Perhaps the most important consequence of this organization is that (structural) properties can be *inherited* from superclass to subclass, e.g., the class of employees inherits properties such as name, address, and so on, from the class of persons.

Generalization hierarchies are often referred to as *is-a hierarchies* in AI. The taxonomic organization provided by such hierarchies can lead to models that are understandable and consistent, because the more that classes have in common with each other, the "closer" they are located to each other in the hierarchy. Also, is-a hierarchies can lead to more concise models, since it is sufficient to associate (structural) properties to the most general, applicable class and let inheritance imply the rest.

The premise posed by the Taxis Project, that this simple framework based on abstraction mechanisms can be used effectively to structure models of all kinds and for all parts of these models, does not mean that this framework by itself is a complete, universal, or general-purpose modelling framework, only that certain "epistemological primitives" [Brachman79] form its primary foundations. In fact, as the reader will see in the next section, special-purpose features are used by Taxis for modelling each system aspect, by allowing several types of objects and properties; they are, however, all moulded into this simple representational framework.

3. The Taxis Languages

This section presents a necessarily brief account of some of the descriptive tools of the Taxis design and requirements languages. More details about the design language can be found in [Mylopoulos80b, 80c], [Wong83] and [Nixon83]. The requirements language is further discussed in [Greenspan82a] and [Greenspan84].

3.1. The Taxis Conceptual Model

3.1.1. Objects

As indicated earlier, a fundamental feature of Taxis is that it provides a conceptual framework for modelling the world. The framework is *object-oriented*, i.e. provides objects as the basic building block for constructing semantic descriptions.

All objects are stratified into *classification levels* according to whether they are considered "individuals", called *tokens*, collections of individuals called *classes*, classes of classes called *metaclasses* and so on. The tokens of a class are called its *instances*; similarly, a class is said to be an *instance* of a metaclass.

Simple examples of object tokens are johnSmith (representing a particular person) and 7 (representing the number 7). PERSON is an example of a class, whose instances are tokens that represent persons, such as the token johnSmith; INTEGER is also a class, with integer objects such as 7 as instances.

An example of a metaclass is PERSON-CLASS whose instances are classes of persons such as PERSON, PATIENT, EMPLOYEE, etc.

We will use lower case letters for token identifiers and upper case letters for class and metaclass identifiers. The suffix "-CLASS" will be used for metaclass identifiers.

3.1.2. Properties -- Factual and Structural

Objects can be related to each other through *factual properties*, intended to represent binary relationships in the world. A factual property has three components: a *subject*, an *attribute* and a *value*. To represent the fact that John Smith's age is 26, we use the factual property specified by the triple

<johnSmith, age, 26>

where the subject, attribute and value are respectively johnSmith, age (an identifier) and 26.

Generic information that pertains to all instances of a class (or metaclass) is represented through *structural properties* associated with a (meta)class. For example, the triple

[PERSON, age, AGE-VALUE]

represents the general statement that every person has an age, by specifying that every instance of the class PERSON can have an age factual property whose value comes from the class AGE-VALUE. (Note that we have used square and angular brackets to distinguish the two types of properties.)

It may be helpful to think of a structural property as a function, e.g.

age: PERSON --> AGE-VALUE

whose domain is the property subject and whose range is the property value. Evaluation of the function for an instance of the domain results in accessing the corresponding factual property, e.g.

age(johnSmith) = 26

An important feature of structural properties is that they determine what can and cannot go in a Taxis model. Thus, in general, if C is a (meta)class with structural property [C, a, V], for every instance x of C it must be the case that a(x) is either *undefined* or an instance of V. This we call the *Property Induction Principle* and it is one of several that define a basic structure for models described in the Taxis framework. (Some other principles are those concerning the is-a hierarchy and are discussed below.)

Since classes are themselves objects and instances of metaclasses, they can have factual properties too. For example, the information

"The average age of persons is 33"

or

"There are 200 nurses"

can be represented with the factual properties

<PERSON, average-age, 33>
<NURSE, cardinality, 200>.

The inclusion of metaclasses in the framework allows the property induction principle to apply to these factual properties as well. This dictates the presence of structural properties such as

[PERSON-CLASS, average-age, AGE-VALUE]
[PERSON-CLASS, cardinality, NUMBER]

in a model that wants to talk about the average age or the size of different person groups.

3.1.3. Abstraction Mechanisms

So far we have discussed two of the three abstraction mechanisms supported by the Taxis framework. The classification abstraction is supported by the "instance-of" relation between objects. Moreover, one can think of an object as an aggregate of all of its properties (e.g., think of a person as an aggregate of a name, an address, an age, a social insurance number, etc.) Properties are thus offered in support of the aggregation abstraction. The property induction principle relates those two abstractions in a coherent way.

The third abstraction mechanism, generalization, is supported by the *is-a* relation that can be defined between classes and metaclasses according to their generality/specificity. For example, the classes PATIENT, NURSE and DOCTOR are all specializations (i.e., *is-a* related) to PERSON:

(PATIENT, *is-a*, PERSON)
(NURSE, *is-a*, PERSON)
(DOCTOR, *is-a*, PERSON)

Clearly, *is-a* has to be a partial order relation which defines a hierarchy (*is-a hierarchy*) for classes and metaclasses. Two important features of the *is-a* relation are:

If (C, *is-a*, D) then every instance of C must also be an instance of D

If (C, *is-a*, D) and D has a structural property [D, a, V] then C must also have a structural property [C, a, V'], where V' is a specialization of V, (V', *is-a*, V).

The second feature of *is-a* suggests that since PERSON has an age structural property, DOCTOR must also have an age property

[DOCTOR, age, DOCTOR-AGE-VALUE]

and that DOCTOR-AGE-VALUE is more specialized, as one would expect, that AGE-VALUE, i.e. it has *fewer* instances (say, persons under 18 can't be doctors).

3.2. The Design Level

We now present a few examples of classes and metaclasses from the design level of Taxis, to give the reader at least a feeling for what is involved in using the design language.

Let's begin by defining the metaclass PERSON-CLASS before we proceed to define some of its instances:

```
metaclass PERSON-CLASS with
    association
        average-age: AGE-VALUE;
        cardinality: NUMBER
end {PERSON-CLASS}
```

Each property in a definition falls into some ***property category***. The two structural properties of PERSON-CLASS have been assigned to the ***association*** property category, which means that corresponding factual properties can change their values over time. In general, each structural property belongs to a unique category and this determines what can be done to the values of that property.

Next we define the classes PERSON, NURSE and PATIENT:

```
PERSON-CLASS PERSON with
    characteristic
        si#: SOCIAL-INSURANCE#;
        ohip#: OHIP#;
        name: PERSON-NAME
    association
        age: AGE-VALUE;
        address: ADDRESS
    key
        personKey: (si#)
end {PERSON}
```

```
PERSON-CLASS NURSE is-a PERSON with
    association
        dept: HOSPITAL-DEPARTMENT;
        salary: DOLLAR-AMOUNT;
        supervisor: HEAD-NURSE
end {NURSE}

PERSON-CLASS PATIENT is-a PERSON with
    characteristic
        hosp-ad#: HOSPITAL-ADMISSION#
    association
        ward: HOSPITAL-WARD;
        doctor: DOCTOR
    key:
        patientKey: (hospital-admission#,ohip#)
end {PATIENT}
```

According to the semantics of the language, *characteristic* properties, such as si# (social insurance number) and ohip# (hospital insurance plan number), cannot change value once they are specified. *Key* properties, on the other hand, define one-to-one mappings from a property value (e.g. SOCIAL-INSURANCE#) to a property subject (e.g. PERSON) and are thus analogous to record or tuple keys. Note that a data class may have several keys (e.g. PATIENT, which has two), or none at all. Thus property categories provide built-in constraints on the values a property can take during the lifetime of its subject.

Transaction classes define operations that can be performed on the data classes of a Taxis specification. For example, the transaction ADMIT-PATIENT takes a person as *parameter* and makes it a patient after checking certain conditions (*preconditions*) to make sure that admission is possible:

```
transaction ADMIT-PATIENT (p: PERSON) with
    precondition
        has-ohip?: ohip#(p) not= nothing;
        space?: admissions < max-admissions
    action
        admit: insert p in PATIENT;
        update: admissions <- admissions+1
end {ADMIT-PATIENT}
```

All information relevant to the ADMIT-PATIENT transaction is specified through structural properties of different categories. Thus the *parameters* are defined by the parameter list for the transaction, *precondition* indicates conditions that must be checked before each execution of the transaction. Finally, *action* properties specify the "body" of the transaction, i.e. the actions to be carried out.

Transaction classes, like all other types of classes, can be specialized in terms of the *is-a* relation. For instance, we may want to define a specialization of the ADMIT-PATIENT transaction which admits patients to be operated on and requires special conditions and actions such as making sure the patient has done all necessary tests:

```
transaction ADMIT-SURGERY-PATIENT (p: PERSON)
        is-a ADMIT-PATIENT (p:PERSON) with
    precondition
        tests?: PATIENT-TESTED(p)
end {ADMIT-SURGERY-PATIENT}
```

According to the semantics of the *is-a* relation, the ADMIT-SURGERY-PATIENT transaction inherits all the properties of ADMIT-PATIENT and has an additional *precondition* ("tests?"). We consider this feature of the Taxis framework particularly useful because it helps the designer of an information system, such as a hospital admission system, conceptualize the large number of conditions, rules and regulations and structure his specification accordingly.

We all know that rules and regulations in a social setting, such as a hospital environment, are bound to have exceptions. Taxis provides the designer of an information system with the capability of specifying *exception classes* and to associate those with transaction *preconditions*. If a *precondition* is found to be false when its transaction is called, the corresponding exception class is instantiated and an appropriate *exception handler*, also a transaction, is called as a substitute for the initial transaction call. We view exceptions as an additional abstraction which can help the designer deal with *overabstraction* with respect to one of the other abstraction dimensions. The availability of exceptions encourages the designer to first deal with the "normal" cases, the "rules", and then worry about the things that can go wrong and how they should be dealt with. It must be added here that our exception mechanism is fairly limited and requires extensions in order to handle some of the exceptional situations we have encountered in a hospital setting.

The facilities we have described so far allow the modelling of entities and operations on those entities. An essential component of any social system involves (long term) *processes*, such as the process of dealing with a hospital patient from the moment he enters the hospital until the moment he leaves. The Taxis framework provides *script classes* as a modelling tool for long term processes. A script is a modified petri net consisting of *states* and *transitions*. For example, the following diagram (from [DiMarco83]) outlines a script for patients who need a (heart) pacemaker.

PATIENT-SCRIPT

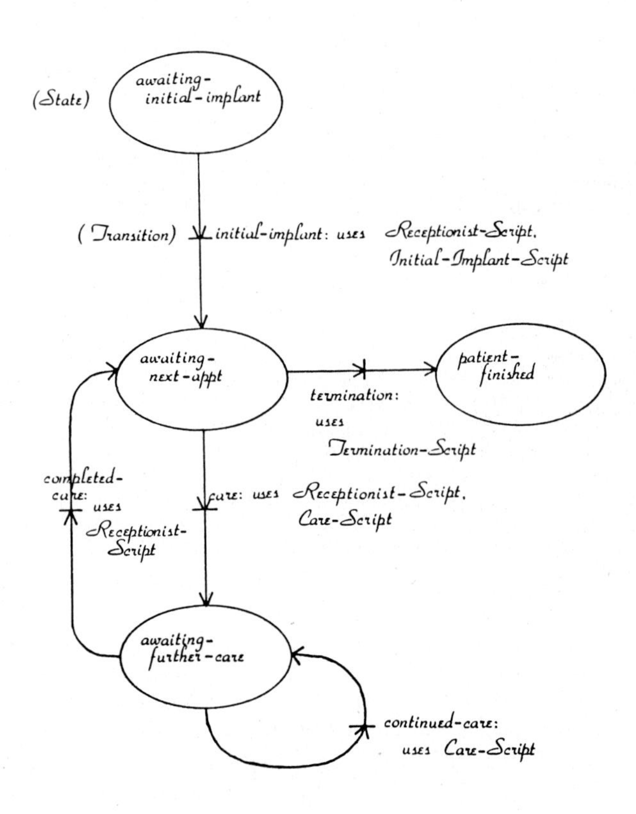

For every pacemaker patient there is an instance of this script which remains active from the time he (or his doctor) decides to have a pacemaker implanted until the time he stops the treatment. Note that the figure only shows states and transitions; it does not show invariant assertions, which can be associated with states and which must be true while a state is active. It also omits the actions associated with each transition, such as accessing or modifying the database. Our script construct, based on [Zisman78] and originally proposed in [Barron80, 82], is further refined in [Pilote83a, 83b]. One of its distinguishing characteristics is that it offers communication facilities (as proposed in [Hoare78]), so that script instances can communicate with each other for synchronization and information-sharing purposes. Another unique characteristic is that scripts are integrated completely into the Taxis framework. Thus, script classes are organized into an is-a hierarchy according to their generality/specificity, have their states and transitions defined in terms of structural properties, and their instances can be accessed through the same facilities used to access instances of, say, data classes.

3.3. The Requirements Specification Level

This level offers, through RML, facilities for *requirements modelling*. A fundamental premise of the Taxis project is that one conceptual framework, such as that adapted here, can be used at the design *and* the requirements specification level. What does change as one moves from one level to the other are the kinds of classes one has at his disposal as he constructs his specification. For the design level these are data, transaction, exception and script classes. For the requirements level, on the other hand, they are *data* ("entity"), *activity* and *assertion* classes. Informally, data classes correspond fairly directly to Taxis data classes. Activity classes are intended to model both instantaneous actions and long-term events and correspond to transactions and scripts. Finally, assertions are logical formulas making statements about the world, the rest of the requirements specification, or the relation between the two. Assertions serve as property values in several useful property categories, and are meant to replace expressions and statements of the design language, and make the requirements level less procedural, as one might have expected. We'll only present here an example of an activity class for the event of admitting a patient:

```
activity ADMIT-PATIENT with
    input
        p: PERSON
    control
        w: WARD;
        phys: DOCTOR;
        consulting-phys: DOCTOR
    output
        pt: PATIENT
    triggered-by
        arr: ARRIVAL(p)
    precondition
        already-in?: not(p instance-of PATIENT);
        ohip?: not (ohip#(p) = undefined);
        space?: cardinality(PATIENT) < maxpatient
    postcondition
        admitted?: IN-HOSPITAL(p);
        patient?: (pt = p) and (p instance-of PATIENT)
    part
        increment: INCREMENT(cardinality(PATIENT));
        put: CHOOSE-WARD(p,w,phys,consulting-phys)
end {ADMIT-PATIENT}
```

The activity defined above has as *input* a person, as *output* a patient, and three *control* properties (w, phys and consulting-phys) which specify objects that are accessed but are not modified by the activity. An instance is created whenever the *triggered-by* assertion becomes true.

Preconditions and *postconditions* relate the activity to assertions which must be true at the start and at the end of the activity, respectively. Note that RML preconditions are very different from Taxis preconditions because the latter are checked at the beginning of a transaction execution (raising exceptions when a false precondition condition is encountered). However, in RML, there is no notion of control flow; rather, the RML conditions form part of the *definition* of the activity and state that for activities in this class the preconditions are true at the start time and the postconditions are true at the end time. In this sense, RML abides by the frequently stated maxim: *The requirements should express* **what** *the system* *should do but not* **how**. *Taxis is used to express how.*

Lastly, we note that activities have ***parts,*** whose occurrences form part of the occurrence of the overall activity. There is no built-in assumption about when or if the parts occur with respect to the overall activity, but rather the specifier may give ***constraints*** (using assertions) to express, for example, that one part must happen before another, or that the outputs from, say, two parts are inputs to a third, in which case the inference can be drawn (but need not be explicitly stated) that the third part cannot finish before the first two have produced their outputs.

When an RML activity is realized in Taxis, all of the objects and constraints of the RML specification must be designed into the Taxis program. Clearly, the designers have quite a free rein to impose design decisions, as long as they do not design a Taxis program that violates these constraints. More strongly, the designed system is responsible for ***satisfying*** the RML specification, which is to say that all RML objects must be somehow represented in the Taxis program, which includes maintaining all ***constraints***. We might add that RML activities do not necessarily get translated directly into Taxis scripts or transactions. The information specified by the activity might be allocated in any number of ways to Taxis objects and properties, e.g. RML ***constraints*** may become Taxis ***preconditions***. Similarly, an RML data object may not become a Taxis data class, but might, for example, be implemented as a value returned by a transaction.

4. Status and Prospectus of the Prototype Implementation

Below we present a pictorial view of the current implementation of the Taxis system.

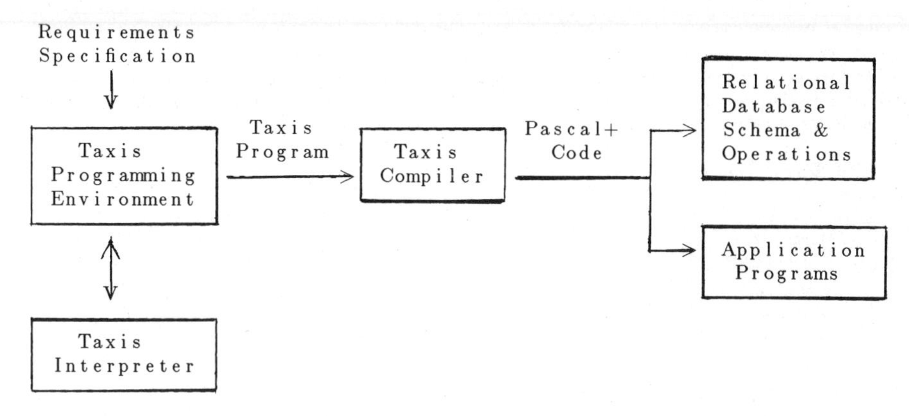

The current implementation only addresses the design level of the software development process pointed out earlier. The Taxis programming environment supports the development of a Taxis program by providing the designer with the necessary tools for constructing and testing prototypes of the program. The Taxis interpreter plays a primary role during this phase, maintaining prototypical data structures and allowing the simulation of algorithms on them. Additional facilities for the management of Taxis programs are also provided within the environment.

Once the development of the Taxis program is accomplished, the program is passed to the compiler, by using an appropriate command of the programming environment. The compiler generates code for a Pascal-like target language augmented with relational database management facilities (e.g. Pascal/R or Modula-R). The generated code includes: a relational database schema for storing data objects, operations for the manipulation of the database, as well as application programs.

4.1. The Programming Environment

The interactive environment was originally designed and implemented by O'Brien [O'Brien82]. It includes a class-oriented editor, whose commands and functionality are centred on Taxis classes; a semantic consistency verifier which insures that Taxis programs conform to the semantic rules of Taxis; and an interpreter and debugger for prototyping. The editor provides the information system designer with facilities to construct, inspect, and modify a Taxis program. The consistency verifier performs various checks to ensure the correctness of the conceptual model being specified. The interpreter simulates execution of Taxis programs, and the debugger assists the designer in validating the model. The design environment also provides various other aids to the user, such as an on-line help facility, a documentation generator, and a way of keeping track of multiple versions of models.

The environment is embedded in a full-screen editor, to make easier and more friendly the interaction between the user and the system. To this end, the set of standard editor commands has been augmented with special functions that operate on the Taxis conceptual schema being built. Menus are provided as a guide in the specification of operations; templates direct the user in the definition of classes.

The functions available in the programming environment fall into three main categories: schema manipulation tools, schema verification tools, and system utilities.

The schema manipulation tools allow the insertion, deletion, modification and display of the objects forming the conceptual schema. The first three function are provided by a class-oriented editor, whose commands and functionality are centred on Taxis classes. Here is an example of the command 'define'. When used, the command causes the following menu to be displayed:

```
# Data Class
# Metaclass
# Transaction Class
# Exception Class
# Finitely Defined Class
# State Class
# Script Class
```

An entry of the menu can be selected by using a special command, identical for the selection from all the menus of the system. After the selection, a template describing the structure of the class to be inserted is made available to the user, who fills the various parts and finally asks the system to accept the definition. The next figure shows the template for the definition of a data class.

```
{ DEFINE dataClass
  and:
  name:
  is-a:
  with
  > characteristics
  > associations
  > assertions
  > keys
  end dataClass }
```

In accepting the definition of a new object, the system performs all the required syntactic checkings to verify that the structure of the object being created conforms to the Taxis syntax. There are several ways of displaying the various portions of the Taxis program being built. The definition of a particular class can be displayed, as well as the offspring of a class within the three hierarchies associated with the abstraction mechanisms provided by the Taxis Data Model.

The schema verification tools include a static semantic analyzer and a simulator of the execution of Taxis programs. The static semantic analyzer can be invoked to verify the consistency of an object definition with respect to the rest of the conceptual schema. The consistency checking is propagated along the instance-of and is-a hierarchies according to the user specification. The program simulator gives the possibility to perform the dynamic analysis of the schema by running Taxis programs. A wide range of debugging options is provided by the system to assist the designer in this phase of the validation of the model. These options can be set up during the simulation of a program, to trace the various stages of the execution.

The most relevant system utilities are help facilities and version control. The user can invoke the help function from different contexts; according to the particular context, the system will reply by displaying the appropriate information. For example, when invoked for an entry of a menu, the help function provides information about the action underlying that entry; when invoked for a field of a template, the help function gives the syntax of the Taxis expression that must be used to fill that field. For version control, the system supplies functions to maintain a number of Taxis programs and their various versions; for each program there is an associated version tree, describing how the different versions of the program are related each other. The user has commands to operate on (create, delete or display) programs as well as on versions of programs.

The full screen editor used for the implementation of the programming environment is Jove, a compact version of Emacs; the interpreter is written in Franz Lisp, and the interface between Lisp and Jove is written in the C programming language. The whole system runs on a VAX 11/780 under the UNIX operating system.

4.2. The Compiler

The translation of data classes produces a relational schema and access code such that the amount of run-time work is independent of the layout (depth, fan-in and fan-out) of the is-a hierarchy. In contrast with other approaches, the compiler design [Nixon83] directly associates all applicable properties with each class. As a result, run-time access to definitions can be made without traversing the is-a hierarchy; this allows rapid maintenance of semantic integrity constraints upon the data objects.

An interesting feature of the compiler is the efficient way in which it implements the association between a transaction call and the transaction definition that applies to that call. The general rule followed by Taxis is that the most specialized version of the called transaction applicable to the parameter value(s) should be associated with each call. All the computations needed to accomplish this policy is done at compile time, so that the run-time system selects from among typically a handful of transaction versions.

Property inheritance is handled in exactly the same way for both transaction and data classes. Strict inheritance is enforced for property specializations; for multiple inheritance the type of the specialized property must be a specialization of the type of each specialized property.

In a system requiring fast interactive responses and long term transaction processing, it is simply unacceptable to compromise the database integrity or to fail to respond to an interactive request. Thus, an important design decision for the run-time system is that there be no fatal errors. These errors are incorporated into the Taxis framework by treating them as *exceptions* raised by the standard operations, which are in turn treated in exactly the same way as user defined transactions are treated. For example, a key constraint violation exception is raised by the insert-object operation. This has the added advantage of allowing the programmer to provide his own handlers to system raised exceptions, if he does not wish to use the default handlers.

As with other efficient implementations of exception handling, there is basically a minimum amount of overhead in providing exception handlers to a called transaction. In fact, the names of the exception handlers associated with a transaction call are treated as a single additional parameter of the call, so that the selection of the appropriate handler at execution time only requires the scan of a typically small list (which is only done if an exception is actually raised).

The is-a hierarchy has distinguished top and bottom classes, *Any* and *None* respectively, both available to the programmer. When *Any* is used as a property value or in a property definition, any object (data object, string, integer) can be the value of that property; besides treating each element of the is-a hierarchy uniformly, this avoids the need for cumbersome "case" statements on classes to ensure type safety when accessing a property value. The amount of checking required for this feature of the language can be greater than the average case, since the search space is potentially the whole is-a hierarchy.

The most specialized class *None* has a single instance, the object "nothing" (which is also an instance of every class), which is used to represent all kinds of null values, like unknown information, missing information, inconsistent information, no value and so on. Special checking is needed when "nothing" is used as a property value; this however can be combined with other checks, which means that the Taxis language allows the use of null values with minimal cost at execution time.

In Taxis there are three different iteration statements corresponding to the three abstraction mechanisms, which provide elegant access to the conceptual schema. For instance,

for each instance i **of** Person **do** <body of the loop>

is used to enumerate all the instances of a class, while

for each subclass c **of** Person **do** <body of the loop>

goes through all the specializations of class Person; and

for each property p **of** Person **do** <body of the loop>

iterates along all the properties defined for the class Person.

The first, and to a lesser extent, the second form of iteration permit fairly standard type checking within the body of the loop. The third one by itself could be checked by repeatedly analyzing the loop body, once for each property; however, the number of possible cases to be checked increases exponentially when the forms are nested. For example,

for each subclass c **of** Person **do**
 for each property p **of** c **do**
  <body of the loop>

Since the language allows arbitrarily nested loops, the compiler cannot in general do all the checking statically. However, checking is done at execution time, in accordance with our requirement that fatal run-time errors be avoided.

4.3. Implementation of Scripts

The design of the Taxis compiler has been extended to handle the execution of scripts and the enforcement of explicit semantic integrity constraints (assertions)[Chung84].

As opposed to transactions, which are mainly concerned with instantaneous activities, scripts model long-term processes, that take place over a long period of time (e.g. days, months or years). Scripts allow the modelling of concurrent processes; their management also requires the system to "wake up" at a particular time or when a set of conditions are met. Additionally, assertions provide the designer with the many benefits of expressing semantic integrity constraints declaratively.

Some interesting problems arise in implementing these facilities within the framework of the Taxis Data Model. As the number of processes grows, it becomes ever more costly to select only those processes whose firing conditions could be met out of all the processes in the system at any one time. Cyclic checking is prohibitively expensive. Similar situations arise in enforcing semantic integrity constraints. Relevant assertions have to be selected for checking after a database update, and those selected assertions then have to be checked by accessing all the data concerned with them.

Chung has proposed an efficient algorithm which ensures that invariants are satisfied in every database state. Similarities between invariant expressions and triggering conditions, in turn, make it possible to apply the same technique to efficient implementation of process management.

5. Towards a New Software Development Methodology

5.1. Taxonomic Programming

A methodology for specification/modelling should provide guidance to its users. At the heart of many software development methodologies lies one or more abstraction mechanisms, which allow us to ignore details at some level, plus a *refinement principle* which provides for the guided and gradual reintroduction of details across the abstraction dimension. The aggregation abstraction forms the core of software design methodologies such as "stepwise refinement" [Wirth71]. Similarly, the "implementation" dimension, the one which we have referred to as "levels", is the basis for approaches such as the abstract machine approach [Dijkstra72]. The generalization abstraction has not been exploited very much in Software Engineering (although Simula does have classes with a rudimentary inheritance mechanism). Yet, it is our contention that it is an invaluable organizational tool for system description in general, and for requirements modelling in particular.

The main idea of specification guided by generalization is that a model can be constructed by modelling first the most general classes, and then proceeding to more specialized classes. For example, in modelling a hospital world, one might consider first the concepts of patient, doctor, admission, treatment, etc. Later, the modeller can differentiate between child patients, heart patients, internists and surgeons, surgical and medical treatments, etc. At each step, only the information (properties) appropriate to that level are specified.

Generalization is the appropriate principle to exploit when the difficulty of modelling is due to a large number of details rather than due to the algorithmic complexity of the system/world; a hierarchy of classes organized along this dimension provides a convenient structure for distributing information (expressed uniformly as properties in RML) and associating it where it most naturally belongs. Such *stepwise refinement by specialization* [Borgida82], [Borgida84] is orthogonal and complementary to the more usual "stepwise refinement by aggregation", whose main effect is to decompose complex situations into a number of less complex ones. Both kinds of refinement are orthogonal and complementary to a third dimension, the progression from "world-oriented" specifications to specifications of a more and more completely implemented system.

5.2. Requirements Modelling

5.2.1. Relationship of RML to Taxis

Since RML is based on the same organizational principles as Taxis, it can also exploit stepwise refinement by specialization. Using RML for requirements definition should be a much easier task than going directly to a design model in Taxis, since Taxis models involve more programming details, given an RML requirements model, the transition to Taxis should be relatively straightforward, since their underlying structure is based on the same framework [Greenspan82b]. Further work is needed to work out a systematic way to add "design decisions" that are needed to derive a Taxis program from an RML model.

5.2.2. Relationship of RML to SADT

The difficulty of building a high-level requirements specification as in RML should not be understated. In the initial stages of requirements definition, all of the parties involved are faced with the problem of deciding what concepts and phenomena are relevant to the situation at hand, agreeing on terminology, and conveying their "mental models" of the situation to each other. Whatever the application domain (e.g., medical, tax law, manufacturing), the knowledge about that application needs to be understood, documented, and communicated.

We propose, therefore, that requirements be defined in two steps:

The first would use a language for ***structured analysis*** such as SADT [Ross77b], in which terms are introduced in an organized way.

The second would use RML for *semantic modelling,* which gives definitions of the semantics of the concepts introduced in the first step.

In [Greenspan84], the connection between SADT and RML is made. SADT was chosen because it has been widely applied and commercially successful, and its simplicity and symmetry imply that deeper principles are at work.

SADT provides a way of introducing concepts/terms into the requirements specification by a process of stepwise decomposition (expanding a concept "box" into a "diagram" containing several interconnected boxes). The result is a hierarchically organized structure of interrelated terms. SADT is founded on a notion of "system" (data and activity aspects inter-related by input, output, and control relationships), so an SADT model may itself be used as a model of the world/system. The SADT diagrams provide a "structured lexicon", a sort of roadmap to guide the RML modelling process.

However, the interpretation of SADT diagrams is dependent on the meaning of words and phrases of the "embedded" language, usually a natural language such as English, that is used to label the boxes and arrows, and additional information about what the diagram means appears in an accompanying natural language narration for each diagram.

RML is used to formalize this information. Classes and properties are defined to correspond to the terms introduced in the SADT diagrams. For each feature of a diagram (box, node, arrow, nodes where arrows split and join), a concept or constraint is defined in RML. The specifier decides on the precise meanings of the words in the SADT model and writes them down in RML. The semantic relationships expressed in the RML model are constrained by the connectivity of the SADT diagram from which it is derived.

RML was designed with SADT in mind, so that data and activity concepts in RML can be used to give the semantics of SADT data and activity boxes, and the various uses of assertions in RML provide the expressive power to express properties of data and activity and any other constraints.

5.3. Prototyping

The ability to quickly get a prototype system from a specification can be extremely useful for seeing if the designed system specifies the intended behaviour and to get feedback from users. Prototypes can be especially helpful for determining what the user really wants. The Taxis interactive environment already provides some facilities that support prototyping, for interpreting Taxis designs, and our future implementation plans are consistent with this philosophy.

5.4. Experience

Taxis was used for describing a medical information system for the Pacemaker Centre at the Toronto General Hospital [DiMarco83]. The system keeps track of patients who have received a cardiac pacemaker. Its job includes keeping accurate medical records, monitoring the patients' status, analyzing patient data, and providing reports. This system is a good example of a system that has a large number of details, and a critical need for data integrity. The semantic component of Taxis was used to describe patients, pacemakers, and medical data, and scripts were used to model the progress of the patients through various activities such as several kinds of assessments and treatments, scheduling and administrative matters.

The most difficult part proved to be the acquisition of knowledge from the hospital staff. No specific method was used for this task, and the experience underscored the need for a requirements modelling step to support this task before Taxis design was undertaken. Once the pacemaker centre operations were well understood, programming in Taxis was relatively effortless. Furthermore, the structure of the Taxis program strongly reflects the actual life-cycles of patients and pacemakers.

Taxis has also been used to describe a medical information system for managing clinical trials [Buchan82a, Buchan82b]. A clinical trial is a controlled experiment in which groups of subjects receive medical treatment and the cause and effect relationship between the intervention and its outcome is investigated. The system keeps track of the subjects, the procedures, and experimental data; it is used to promote both the safety of the patients and adherence to the trial protocol.

Clinical trials prove an interesting application area for Taxis [Buchan82b]:

> There is an intrinsic simplicity and similarity in clinical trials, despite their apparent overwhelming variety. There are prevention and intervention trials; small trials of a few subjects, and large multicentre trials involving tens of hundreds of institutions and thousands of patients. They differ in such details as inclusion or exclusion criteria for the acceptance or rejection of patients into the trial; methods for allocating patients to experimental and control groups; the many types of treatment, investigation and data collection steps that make up the protocol; and so on. But their basic processes, in investigating the cause and effect relationship between some intervention and its outcome, are the same.

It is necessary to capture and organize these concepts in a way that highlights the commonality between clinical trials and suppresses the differences. The abstraction facilities are an aid in this direction, because specialized concepts appear under more general ones in the is-a hierarchy, so that the properties that are the most widely shared come early in the description and more specialized information comes later. Although Taxis has procedural exception-handling, the clinical trial scenario points out the need for additional work on the representation of "exceptional objects", such as one that is to be considered an instance of a class in spite of the fact that it contradicts some property, as in [Lesperance80].

Given a language such as Taxis, one can consider the possibility of developing *generic systems* that can be tailored to a particular situation by specialization and the specification of exceptions.

6. Conclusions and Directions for Future Research

The Taxis Project has designed and implemented a variety of languages and tools for requirements and design drawing ideas from Knowledge Representation. Software engineering is viewed as the construction of a series of models, starting with a world-oriented requirements model (expressed in SADT or RML), then a Taxis design model, and ultimately a completely implemented system. Other work in the same spirit includes [Bubenko81, Mittermeir80, Roussopoulos79, Wilson79, Yeh80].

Current work on the Taxis language is primarily concerned with the completion of a prototype implementation, along the lines described in section 4. This implementation will not include scripts and is expected to be completed by Fall 1986. A second direction of further work for the Taxis language involves the extension of the language to allow the design of form-based user interfaces for the information systems developed through Taxis [Coles86]. At the same time, we are beginning to investigate formally the problems that are raised when a Taxis-like semantic data model is implemented using database management techniques. [Meghini86] describes some initial results in this area, which include a study of the possible logical schemata that might be generated from a conceptual schema, along with a discussion of their advantages and disadvantages. Other problems to be dealt with on this topic include query optimization for semantic data models, concurrency control, transaction processing, security and distribution. Preliminary work by Meghini suggests considerable carry-over of database results to semantic data models.

Turning to requirements modelling, we are working on an extended version of RML (currently called CML) which includes several features that are useful for requirements and not available in RML. Firstly, CML uses a more powerful representational framework which treats objects and links uniformly. This framework is shown to make CML extensible in ways that RML was not, thus leading to a more compact language which nevertheless has (at least) the expressive power of RML. Secondly, CML offers facilities for the specification of exceptions to rules within a particular context. For instance, one can state that

> "Humans live up to 140 years"

but

> "The above rule doesn't apply to the characters of a play"[8],

In addition, CML allows one to talk about the objects in the specification and their referents in the world being modelled. For example, one can express in CML the closed world assumption or an extended closed world assumption of the form

> "The knowledge base finds out about any student registered in the computer science program within a week from the time of his registration"

Yet another feature of CML is that it allows one to talk about the history of the world being modelled as well as particulars about our knowledge of that history (e.g., when did the knowledge base find out something, who provided the information, etc.). [Stanley86] presents a comprehensive account of the language as well as a formal semantics.

We have discussed how and why Knowledge Representation can influence the development of modelling or functional specification languages. Explicit in our arguments is the assumption that software development requires several linguistic levels, some for requirements modelling, others for functional specifications and still others for implementation. What implications does such an influence have on the environments offered for building specifications? Well, to start with, each linguistic level used in a software development process needs an environment. Those for the more procedural levels can offer typical facilities such as special purpose editors, interpreters, tracing and debugging packages, version control and the like, as discussed in section 4.1. The environments for the less procedural levels need reasoning facilities so that a user can probe a knowledge base to see if it is consistent with his expectations. A second type of facility needed for a software development environment supporting several linguistic levels is intended to make it possible to generate lower level (and more procedural) specifications from higher level (and more declarative) ones. Depending on the nature of the two levels, it may be possible to have a compiler that handles this job. Alternatively, the environment may provide facilities for the interactive generation of the lower level specification from the higher level one. These facilities could include expert system features so that the environment plays the role of an active assistant rather than a passive book-keeper in the generation of a specification. A third desirable facility involves the maintenance and management of multiple specifications for a particular software system corresponding to the different linguistic levels supported by the environment. Such a facility would allow a user to maintain a requirements specification, a design specification and an implementation specification of his software along with information on how parts of one specification relate to parts of the others. With such a setup, it is possible to determine how changes of the specification at one level affect the specifications at the other levels. The research project outlined in [Jarke86] focuses on an environment intended to provide all three facilities mentioned above. It is fair to add that there are scores of research issues to be addressed in realizing an environment of the type advocated here and that a research programme addressing such issues can only be described as long-term.

Two final comments on the bottlenecks encountered in applying Knowledge Representation to software development. The first concerns the lack of research results tying Knowledge Representation to the rest of Computer Science (so that we can have our modelling cake and be in a position to computationally eat it too!). Implementation of very large knowledge bases, concurrency control, error recovery, query optimization, compilation are just a few of the issues that have been thoroughly studied by Computer Science for programs and databases and need to be re-examined for knowledge bases. Two recent workshops [Brodie86] and [Schmidt86] provide a glimpse of where we are and how far we need to go with respect to these issues. The second, and more fundamental, bottleneck is our understanding of the nature

---

[8] The context here is the play, which neutralizes the rule on human age.

of knowledge. As we improve our understanding of philosophical and psychological issues concerning the processes that enable *us* to memorize and use knowledge, we will be in a position to offer better knowledge representation techniques, and by extension better software development tools. Ultimately though this enterprise is bounded by progress in Philosophy, Psychology and other disciplines of Cognitive Science. *Software development, and the computational environments within which it takes place, ultimately must be linked to how we process information and conceptualize the world around us.*

Acknowledgments We wish to thank all members, past and present, of the Taxis project for their contributions to the research described in this paper.

References

[Alford77]
Alford, M., "A Requirements Engineering Methodology for Real-Time Processing Requirements," in [TSE77], pp. 60-69, 1977.

[Allen81]
Allen, J., "An Interval-Based Representation of Temporal Knowledge", Proceedings IJCAI-81, Vancouver, 1981.

[Barron80]
Barron, J., *Dialogue Organization and Structure for Interactive Information Systems.* M.Sc thesis, Department of Computer Science; also Technical Report CSRG-108, Computer Systems Research Group, University of Toronto, January 1980.

[Barron82]
Barron, J., "Dialogue and Process Design for Interactive Information Systems Using Taxis" *Proceedings, SIGOA Conference on Office Information Systems,* Philadelphia, PA, 21-23 June 1982, *SIGOA Newsletter,* Vol. 3, Nos. 1 and 2, pp. 12-20.

[Borgida82]
Borgida, A., Mylopoulos J. and Wong, H.. K. T., "Methodological and Computer Aids for Interactive Information System Development", in Hans-Jochen Schneider and Anthony I. Wasserman (Eds.), *Automated Tools for Information System Design and Development, Proceedings of the IFIP WG 8.1 Conference on Automated Tools for Information System Design and Development,* New Orleans, LA, 26-28 January 1982. Amsterdam: North-Holland, 1982., pp. 109-124.

[Borgida84]
Borgida, A., Mylopoulos, J. and Wong, H. K. T., "Generalization/Specialization as a Basis for Software Specification", in Brodie, M., Mylopoulos, J. and Schmidt, J. (eds.), *On Conceptual Modelling: Perspectives from Artificial Intelligence, Databases and Programming Languages,* Springer Verlag, 1984, pp. 87-117.

[Borgida86]
Borgida, A., "Conceptual Modelling of Information Systems", in [Brodie86].

[Brachman79]
Brachman, R. J., "On the Epistemological Status of Semantic Networks," in [Findler79], pp. 3-49.

[Brodie84]
Brodie, M. L. "On the Development of Data Models", in Brodie, M., Mylopoulos, J. and Schmidt, J. (eds.), *On Conceptual Modelling: Perspectives from Artificial Intelligence, Databases and Programming Languages,* Springer Verlag, 1984.

[Brodie86]
Brodie, M. and Mylopoulos, J. (eds.), On Knowledge Base Management Systems: Integrating AI and Database Technologies, Springer Verlag, in preparation.

[Bubenko81]
Bubenko, J.A., Jr., "On Concepts and Strategies for Requirements and Information Analysis," SYSLAB Report No. 4, Dept. of Computer Science, Chalmers Univ. of Technology, 1981.

[Buchan82a]
Buchan, I. M., *A Taxis Clinical Trial Management System.*, M.Sc. thesis, Department of Computer Science; also CSRG Technical Note 28, University of Toronto, 1982.

[Buchan82b]
Buchan, I., H. D. Covvey, J. Mylopoulos, C. DiMarco, and E. D. Wigle, "Taxis: A Language for the Development of Clinical Trial Management Systems," *Proc. Sixth Annual Symposium on Computer Applications in Medical Care,* October-November 1982, pp. 742-747.

[Chen76]
Chen, P. P-S., "The Entity-Relationship Model -- Toward a Unified View of Data", ACM Transactions on Database Systems, Vol. 1, No. 1, March 1976, pp. 9-36.

[Chung83]
Chung, K. L., *An Extended Taxis Compiler,* M.Sc. thesis, Department of Computer Science; Also CSRG Technical Note 37, University of Toronto, 1983.

[Codd70]
Codd, E. F., "A Relational Model of Data for Large Shared Data Banks," Comm. ACM, vol. 13, no. 6, 1970.

[Coles86]
Coles, D. C., *A Hierarchical Framework for Electronic Forms in Taxis.* M.Sc. thesis, Department of Computer Science, University of Toronto, in preparation.

[Dahl72]
Dahl, O.-J. and C. A. R. Hoare, "Hierarchical Program Structures", in Dahl, O-J., Dijkstra, E. W. and Hoare, C. A. R. (eds.), *Structured Programming,* Academic Press, 1972.

[Delgrande85]
Delgrande, J. *A Foundational Approach to Conjecture and Knowledge in Knowledge Bases,* Ph.D. thesis, Department of Computer Science, University of Toronto, 1985.

[Dijkstra72]
Dijkstra, E. W., "Structured Programming," in Dahl, O-J., Dijkstra, E. W. and Hoare, C. A. R. (eds.), *Structured Programming,* Academic Press, 1972.

[DiMarco83]
Di Marco, C., *Using TAXIS to Design a Medical Information System.* M.Sc. thesis, Department of Computer Science; also CSRG Technical Note 31, University of Toronto, 1983.

[Findler79]
Findler, N. (editor), *Associative Networks,* Academic Press, 1979.

[Goldberg81]
Goldberg, A., and Robson, BYTE Magazine, Special Issue on Smalltalk, August, 1981.

[Greenspan82a]
Greenspan, S., Mylopoulos, J., and Borgida, A., "Capturing More World Knowledge in the Requirements Specification," *Proc. 6th International Conference on Software Engineering,* Tokyo, 1982, pp. 225-234.

[Greenspan82b]
Greenspan, S., Borgida, A., and Mylopoulos, J., "Principles for Requirements and Design Languages: The Taxis Project," *Proceedings, International Symposium on Current Issues of Requirements Engineering Environments*, Kyoto, Japan, Sept. 1982. North-Holland and Ohm, Publ.

[Greenspan84]
Greenspan, S. J., *Requirements Modeling: A Knowledge Representation Approach to Requirements Definition,* Ph.D. thesis, Department of Computer Science; also Technical Report CSRG-155, University of Toronto, 1984.

[Hehner85]
Hehner, E. C. R., "Predicative Programming", Comm. ACM, February 1984.

[Hoare78]
Hoare, C. A. R., "Communicating Sequential Processes," Comm. ACM, August 1978, pp. 666-677.

[Jackson83]
Jackson, M., *System Development,* Prentice-Hall, 1983.

[Jarke86]
Jarke, M., Mylopoulos, J., Schmidt, J. and Vassiliou, Y., "Towards a Knowledge Based Software Development Environment", in [Schmidt86].

[Jensen74]
Jensen K. and Wirth, N., *PASCAL User Manual and Report,* (2nd ed.). New York: Springer-Verlag, 1974.

[Konolige85]
Konolige, K., "A Computational Theory of Belief Introspection", Proceedings IJCAI-85, 502-508, Los Angeles, 1985.

[Lesperance80]
Lesperance, Y., "Handling Exceptional Conditions in PSN," in *Proc. 3rd Biennial Conference of the Canadian Society for Computational Studies of Intelligence,* Victoria, B.C., May 1980.

[Levesque74]
Leveque, H. J., "The Logic of Incomplete Knowledge Bases", in Brodie, M., Mylopoulos, J. and Schmidt, J. (eds.), *On Conceptual Modelling: Perspectives from Artificial Intelligence, Databases and Programming Languages*, Springer Verlag, 1984, pp. 165-189.

[Levesque79]
Levesque, H. J., and J. Mylopoulos, "A Procedural Approach to Semantic Networks," in [Findler79], pp. 93-120.

[Lum79]
Lum, V., S. Ghosh, M. Schkolnick, D. Jefferson, S. Su, J. Fry, T. Teory, and B. Yao, *1978 New Orleans Data Base Design Workshop Report,* IBM Research Report RJ2554, San Jose, July 1979.

[McLeod80]
McLeod, D., and J. M. Smith, "Abstraction in Databases," *Workshop on Data Abstraction, Databases, and Conceptual Modelling,* Pingree Park, Colorado, 23-26 June 1980.

[Meghini86]
Meghini, C., Unpublished notes.

[Mittermeir80]
Mittermeir, R. T., "Application of Database Design Concepts to Software Requirements Analysis," Proc. 5th European Meeting on Cybernetics and Systems Research, Vienna, April 1980, Hemisphere Publ.

[Mylopoulos80a]
Mylopoulos, J., "An Overview of Knowledge Representation," *Proceedings of the Workshop on Data Abstraction, Databases, and Conceptual Modelling,* Pingree Park, Colorado, 23-26 June 1980. *SIGPLAN Notices,* Vol. 16, No. 1, June 1981, pp. 5-12.

[Mylopoulos80b]
Mylopoulos, M., P. A. Bernstein, and H. K. T. Wong, "A Language Facility for Designing Interactive Database-Intensive Applications," *ACM Transactions on Database Systems,* Volume 5, Number 2, June 1980, pp. 185-207.

[Mylopoulos80c]
Mylopoulos, J., and H. K. T. Wong, "Some Features of the Taxis Model," *Sixth International Conference on Very Large Data Bases,* 1-3 October 1980, pp. 399-410.

[Nixon83]
Nixon, B. A., *A Taxis Compiler.* M.Sc. thesis, Department of Computer Science; also CSRG Technical Note 33, University of Toronto, 1983.

[O'Brien82]
O'Brien, P. D., *Taxied: An Integrated Interactive Design Environment for TAXIS.* M.Sc. thesis, Department of Computer Science; also CSRG Technical Note 29, University of Toronto, 1982.

[Park85]
Park, S. G., *TAXIED-e: Automation of Scripts and User Interface in an Integrated Interactive Design Environment for Taxis.* M.Sc. thesis, Department of Computer Science; also Technical Note CSRI-39, Computer Systems Research Institute, University of Toronto, 1985.

[Pilote83a]
Pilote, M., *A Framework for the Design of Linguistic User Interfaces.* Ph.D. thesis, Department of Computer Science; also CSRG Technical Note 32, University of Toronto, 1983.

[Pilote83b]
Pilote, M., "A Programming Language Framework for Designing User Interfaces," *Proc. of the SIGPLAN '83 Symposium on Programming Languages Issues in Software Systems. SIGPLAN Notices,* Vol. 18, No. 6, June 1983, pp. 118-136.

[Quillian68]
Quillian, M. R., "Semantic Memory", in M. Minsky (ed.): *Semantic Information Processing,* MIT Press, 1968, pp. 227-270.

[Ross77a]
Ross, D. T., and K. E. Schoman, "Structured Analysis for Requirements Definition," in [TSE77], pp. 6-15.

[Ross77b]
Ross, D. T., "Structured Analysis(SA): A Language for Communicating Ideas," in [TSE77], pp. 16-34.

[Roussopoulos79]
Roussopoulos, N., "CSDL: A Conceptual Schema Definition Language For the Design of Data Base Applications," *IEEE Trans. on Software Engineering,* Vol. SE-5, No. 5, Sept. 1979.

[Schmidt80]
Joachim W. Schmidt and Manuel Mall, *Pascal/R Report.* Bericht Nr. 66, Fachbereich Informatik, Universitaet Hamburg, Jan. 1980.

[Schmidt86]
Schmidt, J. and Thanos, C. (eds.), *On Knowledge Base Management Systems,* Springer Verlag, (in preparation).

[Shibahara85]
Shibahara, T., *On Using Causal Knowledge to Recognize Vital Signals: A Study of Knowledge-Based Interpretation of Arrhythmias*, Ph.D. thesis, Department of Computer Science, University of Toronto, 1985.

[Smith77]
Smith, J., and D. C. P. Smith, "Database Abstractions: Aggregation and Generalization," *ACM Transactions on Database Systems,* Volume 2, Number 2, June 1977, pp. 105-133.

[SRS79]
Proc. of the Conference on Specifications for Reliable Software, Boston, 1979.

[Stanley86]
Stanley, M., *CML: A Knowledge Representation Language with Application to Requirements Modelling,* M.Sc. thesis, Department of Computer Science, University of Toronto (in preparation).

[Teichroew77]
Teichroew, D., and E. A. Hershey, III, "PSL/PSA: A Computer-Aided Technique for Structured Documentation and Analysis of Information Processing Systems," in [TSE77], pp. 41-48.

[TSE77]
IEEE Transactions on Software Engineering, Volume SE-3, Number 1, January 1977.

[Tsichritzis81]
Tsichritzis, D., and F. Lochovsky, *Data Models.* Englewood Cliffs, NJ: Prentice-Hall, 1982.

[Wasserman77]
Wasserman, A. I., *Procedure-Oriented Exception-Handling.* Technical Report 27, Medical Information Science, University of California, San Francisco, Feb., 1977.

[Wilson79]
Wilson, Max L., "A semantics-Based Approach to Requirements Analysis and System Design," *Proc. COMPSAC 79,* November 1979, pp. 107-112.

[Wirth71]
Wirth, N., "Program Development by Stepwise Refinement," *Comm. ACM,* Volume 14, Number 4, April 1971, pp. 221-227.

[Wong83]
Wong, H. K. T., *Design and Verification of Interactive Information Systems Using TAXIS,* Ph. D. Dissertation, University of Toronto, 1983.

[Yeh80]
Yeh, R., Zave, P., Conn, A., and Cole, G., Software requirements: a report on the state-of-the-art, Univ. of Maryland, TR-949, Oct. 1980.

[Zisman78]
Zisman, M. D., "Use of Production Systems for Modeling Concurrent Processes", In D. A. Waterman and Frederick Hayes-Roth (Eds.), *Pattern-Directed Inference Systems,* New York: Academic Press, 1978, pp. 53-68.

Design and Applications of Expert Systems

Bernd Radig

Fachbereich Informatik der Universität Hamburg

Abstract

The successful application of expert systems depends much more upon a systematic, carefully tailored, tool supported development process than conventional software projects. This contribution analyses the requirements from application areas and the obstacles which hinder to satisfy the demand for expert systems. The architecture of expert systems is presented to identify necessary subtask to follow during the development. The roles of a life-cycle model and of participants in the design process are identified. The need is pointed out to extend development methodologies so that they cover also the building of embedded, distributed, and real time expert systems.

Application

Expert systems are invented to support human problem solving. Some kind of knowledge is required to arrive at the solution, if it exists. In conventional data processing, expertise is needed by the systems engineers which enables them to find an algorithm which in turn implements the solution of a task. In contrast, expert systems are confronted with a task and are requested to apply their knowledge to find a particular one or all solutions to the problem.

From an expert's point of view, an expert system

◇ is a tool which helps an expert to analyse and codify domain knowledge,
◇ makes his past experience and knowledge available for professional evaluation,
◇ gives the expert insights into his knowledge area,
◇ allows a pooling of knowledge sources from different domains.

From the user's point of view, different roles of expert systems can be identified:

◇ The system replaces or multiplies a human expert, it is in a limited way a human expert imitator. In this case, the user does not posess the knowledge, expertise, or skills which are needed in the application domain in order to solve a problem. The user's knowledge is a subset of the system's knowledge, just enough to formulate and present the problem as well as to understand and to apply the proposed solution. To give an example, configuration of computer hardware or operating systems can be done more reliable by experts systems than by engineers [McDermott82b], [Haugeneder+85].

◇ The system acts as a colleague. It helps the user to check his own reasoning, reminds him of alternative solutions, or guides him on the solution path. Therefore, the domain knowledge of the system differs only slightly from that of the user. XSEL is such a system which helps a salesperson to get a precise enough impression what kind of a computer systems could match his client's needs [McDermott82a].

◇ The system acts as an assistant. It frees the user from boring elaboration of his hypothetical solutions or offers to him support in areas which are not central for the problem and might stem from other domains. An example is mathematics where MACSYMA performs symbolic manipulation of algebraic expressions, including symbolic integration, canonical simplification, or solution of equations [Martin+Fateman71].

◇ The system substitutes a human in cases where humans are - in principle - not able to handle a problem solving task. Processing of sensory data [Nagel85], [Riseman+Hanson85] or interpretation of process data in real time [Turner86] is an example of such cases.

Expert systems are built to increase human capabilities of problem solving in several directions:

◇ They might be faster. This is either a necessary requirement such as in real time expert systems, or it is comfortable to have a solution of some subproblem quickly available.

◇ They do not overlook relevant information - if the knowledge sources correctly direct the system to utilize relevant information. They neither forget to apply an internal knowledge source nor do they miss problem data supplied by the outside world. This ability is strongly required in monitoring applications, e.g. when computer operators monitor and control an operating system [Griesmer+84].

◇ They are able to handle simultaneously a vast amount of information. They make an optimal solution available when human experts are overloaded by details. They don't miss e.g. a high revenue opportunity.

◇ They do not panic. Even in cases when humans fail to monitor effectively all available information, expert systems are able to select from the information flow relevant parts.

◇ They do not make the same mistake twice - if the cause of the mistake can be traced down to the responsible knowledge chunks. But it can happen that they make mistakes, since human incomplete, erraneous expertise is mapped into the system's knowledge base.

◇ They are able to resolve conflicting constraints in a consistent, reproducible way. In decision making, finding an optimal solution in the presence of conflicts is only half the task. The other half is to argue for the proposed solution, especially when the implicated subsequent actions are to be performed by humans. Job-shop scheduling, as an example, is in this category of tasks [Fox+Smith84].

◇ They are able to follow long and complicated reasoning chains, where humans tend to get lost. Theorem proving or formal program verification are activities which are expected to be tailored for expert systems application.

Despite their potential usefulness, there are some obstacles which stand against a wide spread use of expert systems. Building expert systems [Hayes-Roth+83] is a technically difficult area where no standard architectures and methodologies are readily available to support the design and maintenance process. Development of expert systems is costly and is successful only if the problem domain is carefully selected. Among these obstacles are

◇ knowledge acquisition: The task is to map relevant pieces of knowledge from the application domain into the system's knowledge base. This knowledge can be acquired from human experts, published research reports, textbooks, case studies etc. There are no methodologies available consolidated enough to aid the systematic and economical collection and transfer of expertise.

◇ System architecture: The architecture of the expert system should reflect the characteristics of the problem class to be solved. There is some experience that e.g. for a diagnostic system a MYCIN like structure is sufficient. But again, a systematic guide what type of an architecture to use for what types of application is not available., rather, the choice is dictated by the tools, shells and languages which are at hand or which the system builders are acquainted with.

◇ User interface: An intelligent system should exhibit an intelligent behaviour towards the user, proving its skills not only by coming to a solution for the user's problem but also by giving explanations and arguments. The ability to interact with the user in his natural language is regarded as an essential step to increase the acceptance of expert systems. Despite the progress made so far, the mere analysis and generation of written language is not sufficient. The focus of research shifts to problems of partner modelling, non-grammatical phrases, indirect speech acts etc. [Wahlster86].

Besides improving natural language understanding and generation, a human being usually interacts with his environment not only by writing and reading. Understanding and generating continous speech is a challenging task in its own right. Of similar complexity is the understanding and generation of images. An expert system, equipped with a fully-fledged visual, acoustical, and textual interface would need computing power beyond our present technological possibilities to stand the requirements of a communicative user in real time.

◇ Knowledge maintenance: Test and evaluation is complicated. On one hand, it is not guaranteed that the knowledge, acquired from the experts, is correct. The human experts are able to check the agreement between their solution and that of the expert system. If the knowledge engineer has attempted to emulate the expert's problem solving strategy, and the chain of conclusions is not too long, and the amount of data involved is not too great, the expert might be able to follow and evaluate the system's solution path.

On the other hand, case studies or other kind of test input is not always available. Here the same techniques are applicable which are utilized when acquiring expertise. The knowledge base should be corrected when the system makes a mistake. If mistakes are too expensive in the real world, a (correctly) simulated environment can be helpful. In developing expert system, it is not well understood how to implement what is characteristical for a good expert: to be motivated, to gather experience, and to update his knowledge, and to exercise critical introspection in order to define his range of competence. If the system's knowledge can become irrelevant for the problem domain then, in the present situation, only the supervision of the knowledge base by human experts will prevent the system to become fossilised.

◇ Delegation of responsibility: Really employing an expert system means that the user trusts the system the way he would trust his assistant, colleague or another expert. This is not a completely new phenomenon in our computer age. We are acquainted to trust in our daily life a huge amount of results produced by complex algorithms, executed by digital computers. But expert systems are still in their infancy, based on the new, not completely understood technology of Artificial Intelligence.

Architecture

A characteristic of knowledge-based systems is the strict separation of the explicit knowledge base from the modules which acquire and apply this knowledge. A very coarse model of system shells such as EMYCIN [VanMelle+84] reflects this characteristic (Fig.1).

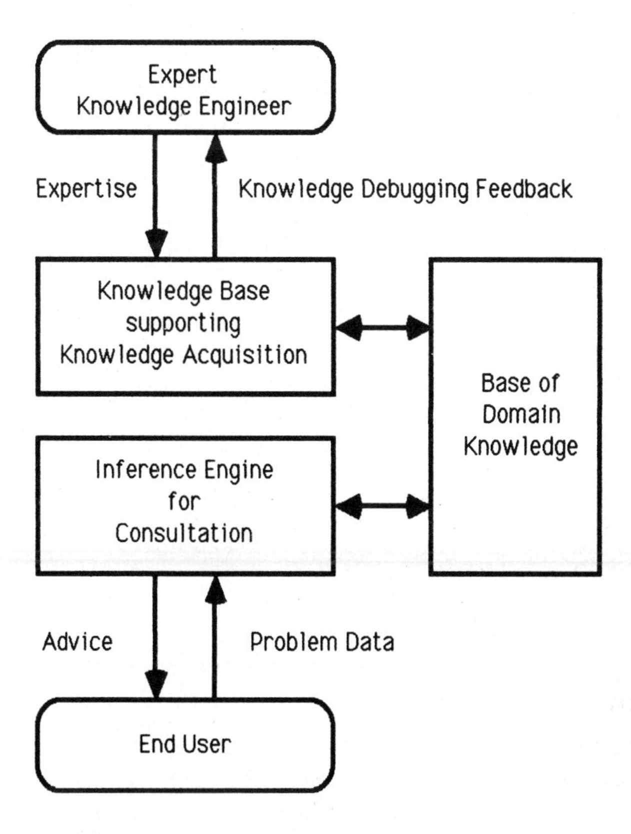

Fig. 1: Coarse View on EMYCIN

EMYCIN supports the knowledge engineer to build the knowledge base, providing him with preformulated mechanisms to formulate rules and to organize facts. It supports a backward chaining inference strategy and the attachment of confidence values to enable fuzzy reasoning. It was a step forward to extract from the MYCIN experience the basic architectural and functional principles, stripping off domain dependant knowledge. Some more shells have been constructed by abstraction from special expert systems [Waterman85]. Other tool development starts from a language, problem oriented for representation and reasoning. These tools combine several paradigms such as rule-oriented, logic, object oriented, functional programming with different strategies such as event driven or goal directed exact, approximate, or nonmonotonic reasoning, e.g. LOOPS [Bobrow+Stefik83] or BABYLON [DiPrimio+85]. All kind of frames and semantic networks are widely employed to structure data, giving the system designer built-in concepts such as abstraction, inheritance, and instantiation.

These tools are a step forward in the direction of an efficient implementation of expert systems, but do not support sufficiently the time consuming design phases which have to preceed the implementation. Furthermore, only recently languages and tools are prepared to interface with a complex external world where the system might be embedded in. Fig. 2 shows a more detailed view of an expert system, extending an architecture model from [Raulefs84].

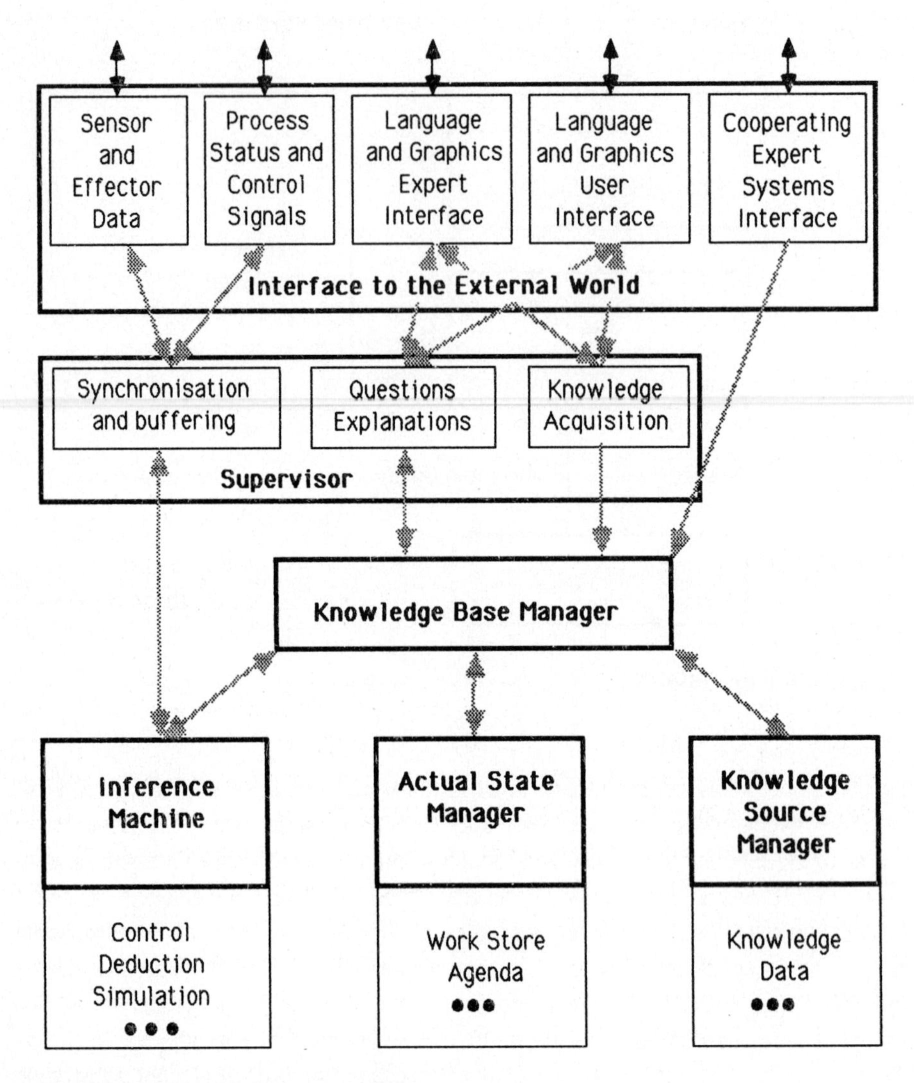

Fig. 2: Architecture of Expert Systems

◇ The inference machine controls and executes all reasoning towards achieving the problem solution. The control system guides deduction or simulation steps. It is resposible to split goals into subgoals or to direct a search procedure. Especially in real time expert systems it is important to reduce the search space by implementing sophisticated search strategies which use heuristic approaches. The knowledge base usually contains some kind of meta-knowledge which helps to selecting the appropriate strategy and heuristics. The control system evaluates the emerging model of a solution and determines which part of it has to be extended next in a goal or data directed manner. It decides which knowledge sources to apply and evaluates the extension to monitor the progress and to check whether a solution has been reached. Efficient techniques for matching relevant knowledge to current problem data must be available as part of the inference machine [Turner86].

◇ The synchronization modul within the general supervisor is responsible to provide the inference machine with problem data arriving from external processes and sensors. In the opposite direction, it provides at the right moment the effectors and processes with control signals. These signals are the vehicle for the expert system changing its external environment physically.

◇ The actual state manager keeps a record of goals and subgoals in its agenda and provides the working store for intermediate results. Sometimes it is integrated in the inference machine, sometimes it appears explicitly as in the case of the blackboard paradigm.

◇ The knowledge source manager organizes the various knowledge sources including factual knowledge. It should be able to employ e.g. a conventional database management system.

◇ The knowledge base manager controls the flow of information between the knowledge sources and the work store. It is responsible to identify appropriate chunks of knowledge according to the agenda of the actual state manager. Additionally, it supervises the maintenance of the knowledge base. Modification of the knowledge base takes place during a knowledge acquisition phase. In a distributed expert system architecture, information may be obtained from cooperating expert systems which solve in parallel appropriate subtasks and whose results either influence the working store and the agenda or are fed directly into the knowledge base. The knowledge base manager should care that the knowledge base is in a consistent state, where consistency is defined according to the specification of the problem domain characteristics. To implement a version history and to control the access rights to the knowledge base is also among its tasks.

◇ The knowledge acquisition module supports the introduction of new expertise or the refinement of already existing knowledge sources. It translates knowledge representation from an external form as supplied by the expert and end user interface to the form as required by the knowledge base manager. It informs the knowledge engineer of consequences of modifications and may inform the knowledge base manager to perform undo or redo actions.

◇ The third module in the expert system supervisor handles questions and explanations. It collects all the information, mostly from the inference machine and the actual state manager, which informs the end user as well as the expert on which path a problem solution is under way and which knowledge sources have been used so far.

◇ Several interfaces connect the expert system with its environment. The end user as well as the expert should be provided with means to communicate using graphics and text. A language interface is simple for a menue driven dialogue. To perform a real natural language dialog is beyond today's possibilities. A graphic oriented browser is extremely helpfull for the knowledge engineer to get information about objects, their attributes and their relationships. A graphics interface is a must when visual information is involved or information can be visualized. Visualization means to see or form a mental image of what is going on in the expert system [Herot+82]. It helps the users in the formation of a clear and correct mental image of the structure and function of the expert system's knowledge and problem solving strategies.

If the expert system controls a process or plans and initiates actions of a robot, it has to collect information and to generate commands. Usually, the process status or the sensor signals are not suited to be directly fed into the inference engine. The signals are condensed, classified and converted to symbolic values. It is not difficult to convert subsequent readings of a temperature sensor into the statement that the temperature is rising with a rate of some degree per second. Much more complicated is the task if the interface has to interpret an image from a visual sensor. To perform here the transformation from an iconic to a symbolic description efficiently and eventually under real time constraints, is a poorly understood subject.

If the expert system is to produce command signals for a process or for an effector, the interfaces usually have to translate from a symbolic expression to a form which corresponds to the requirements of the process or effector control systems. Controlling a manipulator it is a good practise to replace - during the development stage - the effector interface by the graphics interface in order to simulate the plans which the expert system has worked out.

The expert system maintains a model of the static and dynamic world it is existing in and partially of its own behaviour. The module structure of Fig.2 can be viewed from the point what parts of the model are located where.

◇ The knowledge base manager - together with the inference machine, the actual state manager, and the knowledge source manager - models the expert system's knowledge processing. Inference strategies, the agenda, and meta-knowledge describe the problem solution capabilities and progress.

◇ The sensor and effector interface, together with the process interface model the technical process which the expert system has to monitor and to control. They usually contain also data communication facilities with other processors in a distributed system.

◇ Communication facilities with humans are contained in the expert and end user interfaces. They model two man machine interfaces. One for the expert or knowledge engineer utilized during system design and knowledge maintenance. The other for the end user when the expert system is in productive operation.

◇ In some applications it might be reasonable to have a distributed expert system arrangement. For example, in an industrial automation application a design expert system hands its results over to a planning expert which specifies the fabrication process which is in turn online monitored and controlled by a third expert system. The expert system interface maintains the model of the external, cooperating partner systems in so far as the systems are able to communicate and to be synchronized with each other. The model should specify when and how the partner system is expected to react and what to do in the case the partner is down or the communication channel is overloaded or interrupted.

◇ The supervisor modules link the external model with the internal system model. The synchronization model describes the dynamic behaviour of the external process as well as of the reasoning process. It specifies what to do if the real time requirements of the external process do not match the problem solution progress. It knows how to interrupt the inference machine when some unexpected event in the external world occurs.

◇ The question and explanation model reflects the strategy how the human partner is informed about what is contained and going on in the system and its knowledge base. It specifies how to translate a user's questions into a representation which can be understood by the knowledge base manager. In the other direction it defines how the internal information should be presented to the outside world.

◇ The knowledge acquisition modul supports the transfer of expertise by providing a interpretation model of the data which may be entered by the knowledge engineer [Hart85].

Design

Knowledge acquisition: elicitation and interpretation of data is the phase in which the knowledge base is built and maintained. Therefore, it is a permanent activity which accompanies the whole life time of the expert system. In the design phase, the knowledge base interpreter machine is specified, implemented, and tested together with the surrounding modules (Fig.2).

It is well recognized that the design and evolution of knowledge based systems is a complex process which is not properly understood. Nevertheless, the commercial application forces the development of a design methodology [Schachter-Radig85] which can be supported by automatic tools [Breuker+Wielinga85]. Waterman lists common pitfalls and gives some advices how to avoid them [Waterman85]. Major problems are identified in [Breuker+Wielinga]:

◇ It might be not clear from the beginning if the selected domain is apt for the building of an expert system. It is difficult to calculate the costs and to determine the efforts in advance. How much experience and technical support will be required is as well an open question as the availability of toolkits whose built-in formalisms adequately represent the type of knowledge and inferencing needed in the domain.

◇ Knowledge acqusition is difficult. Books and other documents contain mostly the theoretical basis, first principles, or factual data. The application of this kind of support knowledge usually can be only inferred from human experts, observing their expertise in action. This means that the interpretation of verbal data is a crucial step, complicated by the observation that verbal data are incomplete, unstructured, unreliable and contradictory. Interpretation models are not readily available which structure these data into a coherent description and map the description onto some representation formalism.

◇ The developed and validated expert system may fail to be accepted in the intended operational environment. Especially for real time expert systems is it difficult guarantee that preset performance criteria can be satisfied. It might turn out that not only the problem solving competence is not sufficient but also the system's competence to communicate with the environment. Fig. 3 shows a life-cycle similar to that in [Buchanan+83]. A failure due to the mentioned reasons usually means a step back to the initial phase of problem identification.

◇ Expert systems require extensive maintenance. In many domains of expertise, knowledge and practise may change with time. This implies sometimes not only modifications in the implementation but the reactivation of the knowledge organization or even the conceptualization phase. Experience tells that expert systems - prototypes - become unstructured. Then it is hard to identify where modifications and updates should be applied. Simply adding some rules may lead to unforeseen conflicts and may decrease competence as well as performance.

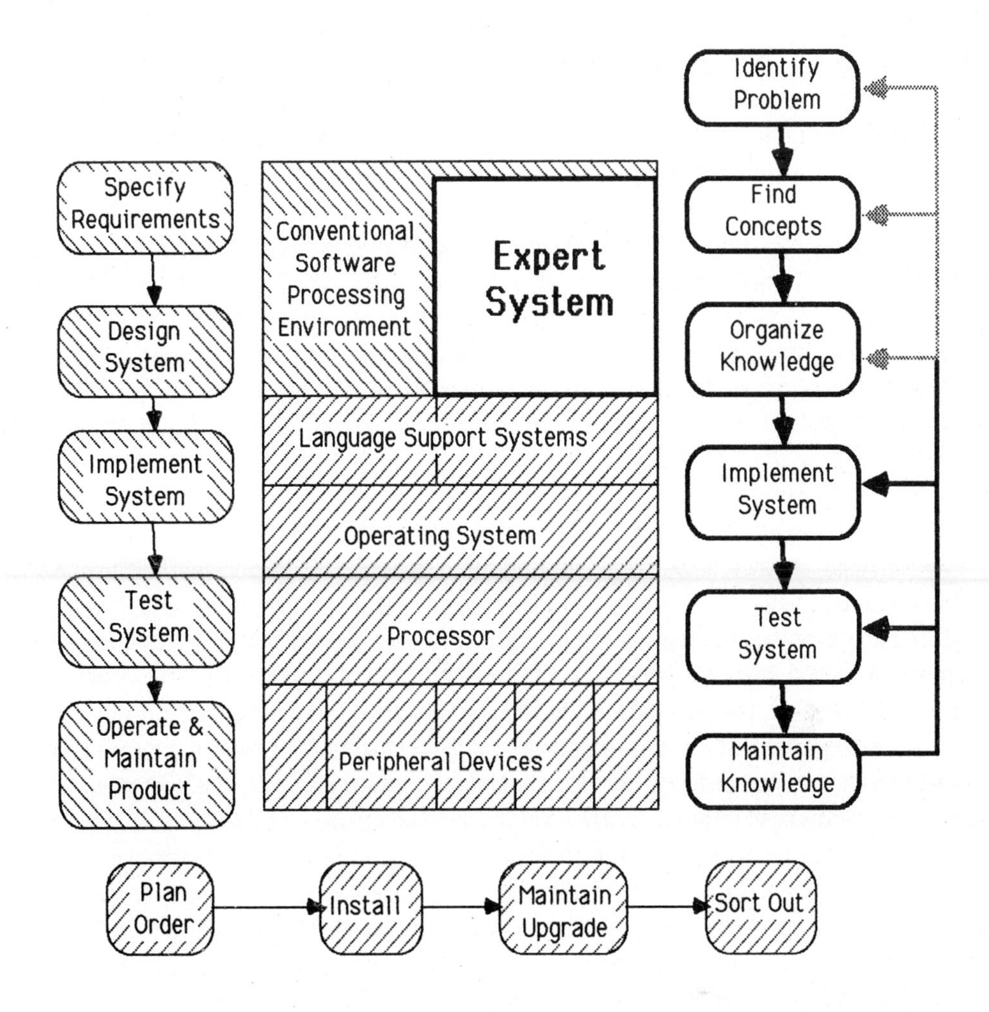

Fig. 3: Life-cycles of Expert Systems, Data Processing Software, and Hardware

Design of expert systems which are not stand-alone but rather embedded in an existing processing environment interacts with the life-cycles of the other components. Fig. 3 illustrates also simplified life-cycles of the hardware and the conventional software environments.

Maintaining and upgrading of processor hardware should not impose essential modifications onto the expert system. The severe problem arises when the hardware is taken out of operation during the life time of the expert system or when a new version of the operating system is somehow incompatible with the old one. The transfer of the expert system to a new computer requires either that the implementation language is portable or has to be adapted to the new hard- and software. To achieve portability of AI-languages and -tools is a problem which waits for a solution. The standardization of the language LISP as CommonLISP and of the toolkit LOOPS as CommonLOOPS are indications that this problem has been recognized.

If the expert system communicates with external, conventional software, the more straight forward development strategy of conventional software interferes with the iterative, fast prototyping strategy of expert system development. The requirement specification may not be clear until the expert system has passed the implementation phase the first time. The interaction and information exchange cannot be tested on the expert system's side until the conventional system has also passed its implementation phase.

It is a challenge to develop a methodology for the creation of software systems and of knowledge based systems. To develop a common methodology which supports the whole life-cycle of knowledge bases, AI-languages and tools, data processing systems and the underlying hardware seems to be a too complex task at the moment. Ad-hoc solutions have to be followed, but they tend to make the development process expensive, unpredictable in terms of time and effort, and unreliable.

An alternative relieves some of the problems. The expert system together with its underlying hardware should be coupled as loosely as possible with the external systems. Progress in standardization of communication protocolls and high speed communication links supports the implementation of distributed systems where one or more of the components can be dedicated to an expert system. The advantages of the object oriented programming paradigm - encapsulation and activation via message passing - may be transferred to the design of complex, interacting, and communicating knowledge based systems.

There is a forth life-cycle involved which influences the design and redesign of an expert system. The external machinery which is under control or supervision of the expert system will change over time. Machines are sorted out, new ones are employed which usually behave different. Minor changes may only require modifications in the interface and synchronization modules (Fig.2). Major changes will initiate a new round in the development cycle (Fig.3).

A life-cycle is only one aspect of the description of the expert system development process. Another aspect identifies the participants in that process and their roles (Fig. 4).

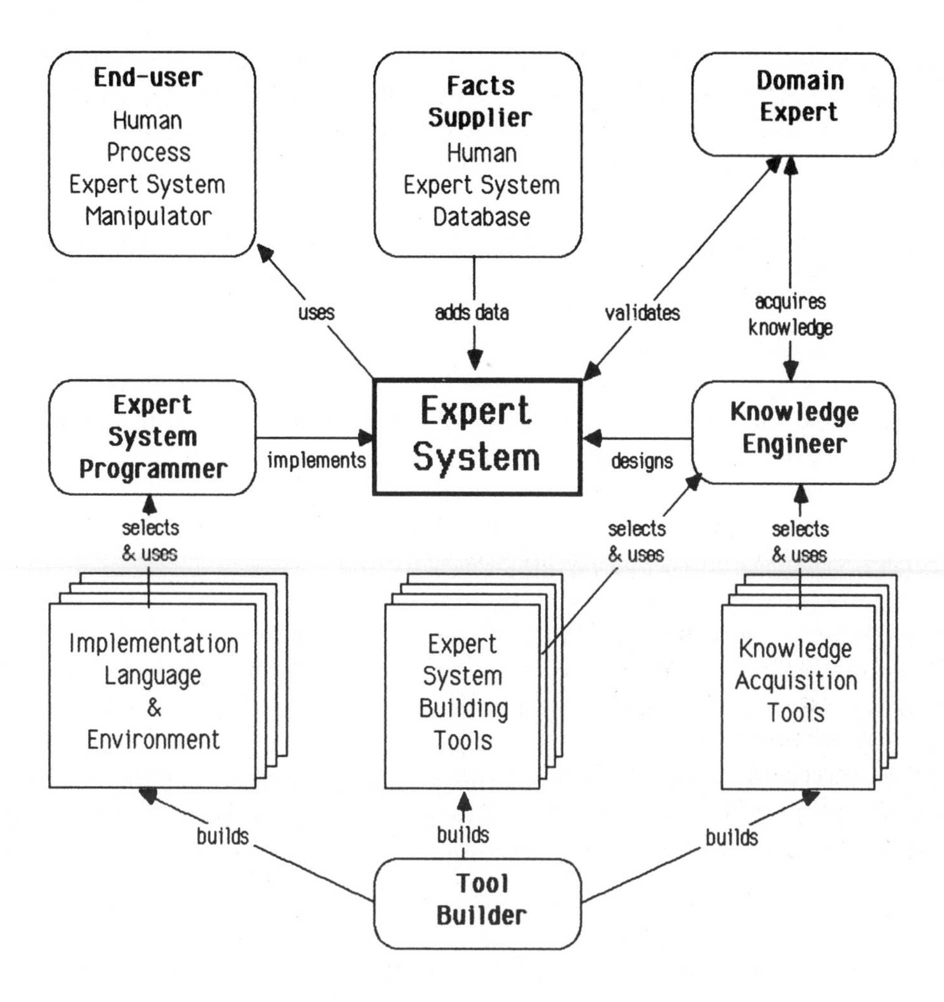

Fig. 4: Participants in the development process

During the problem identification, conceptualization, and knowledge organization phases, the domain expert and the knowledge engineer work closely together. Feasibility analysis and knowledge acquisition are major tasks. Both, domain expert and knowledge engineer, are involved in the test and maintenance phases. But here they should work independent from each other, it might be wise to ask another domain expert to validate the expertise of the system. In any case, the engineer should be equipped from the domain expert with a bunch of case studies which he can use to check the function and performance of the system after each modification.

The knowledge engineer will select among knowledge acquisition tools in order to aid and document the acquisition process. Documentation is of extreme importance in each software development process. But if - in some cases - the source code of an algorithm can be inspected to find out why the function fails, it seems to be more or less impossible to reconstruct the domain knowledge from the knowledge base of an expert system. Furthermore, when the domain knowledge is changing and when their exists no well documented protocoll of the knowledge acquisition activities, the link between the knowledge base and the domain knowledge is lost. Then it is extremely difficult to find out what rules and what inference strategies should be modified in order to reflect the changing expertise.

When the design process iterates in the conceptualization and knowledge organization phase, expert system building tools will aid in developing the first prototype. Here, the knowledge engineer should prefer a tool which might not be the most efficient one but rather offers support for a broad range of concepts for each of the modules from Fig.2. The available tools in their majority support only the implementation of the inference engine as well as structuring and managment of the knowledge base. Some support also an explanation facility.

The task of implementation can be left to an expert system programmer who chooses appropriate implementation languages for the different modules. He is responsible to formulate syntactically and semantically correct rules, to observe the knowledge representation structure, to interface the modules and to construct the interfaces to the external world. More and more AI-languages (PROLOG, LISP, OPS5, etc.) are equipped with call-out interfaces to foreign languages such as C, Pascal, or FORTRAN. For example, OPS5 and Ada are a powerful combination for image understanding expert systems.

The expert system is developed to be used by humans, other processes, other expert systems, or a manipulator and similar kind of machinery. A human end user is always involved. If he is not the one who utilises the problem solution of the expert system - it rather addresses a process - he will play the role of a supervisor, asking the expert system for explanation and argumentation in order to monitor its competence.

The expert system needs problem data as well as the updating of its factual knowledge. Facts may be supplied by human operators, by database systems, or by cooperating expert systems as part of their task.

The domain expert should be available during the phases when the knowledge base has to be updated. This updating may be initiated by the change of expertise in the problem domain or by malfunctions of the system. It is desirable that the domain expert interface and the knowledge acquisition module (Fig.2) are powerful enough to allow the domain expert to change the knowledge base without the help of an knowledge engineer.

When an expert system is delivered and in operation only the end user, facts supplier and occasionally the domain expert should be needed (Fig.5). One consequence is that the expert system might be reimplemented or transfered from a development computer system to a delivery machine which is usually cheaper. This is a necessary prerequisite in such applications where the expertise should be available at the working place of people. Again, portability is required or the commitment to a manufacturers product family which includes compatible development and delivery systems.

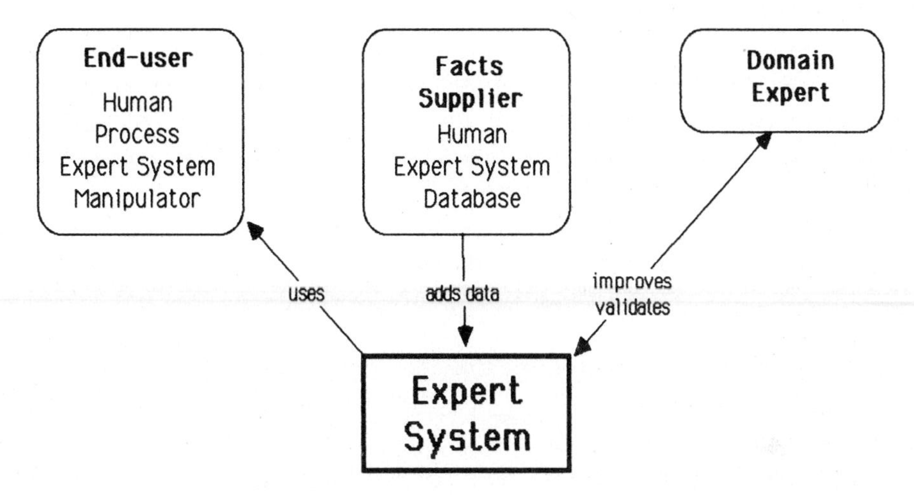

Fig. 5: Delivery System in Operation

Conclusion

Design, implementation, and maintenance of expert systems are still an art. There is a strong presure from commercial, industrial, and scientific users to develop expert systems to help them solving their problems. But still most of the systems are built as feasability studies, some of them found their way as prototypes into some productive environments. Only a few of them do continously useful or profitable work.

The situation will improve when the development process is better understood. Then knowledge engineers will master the design complexity of an expert system embedded in a real environment, communicating with conventional data processing and process control software, and being used by real users. The demand for real time expert systems can be satisfied when either special hardware supports the fast execution of inference steps or when the theory of making decisions under restrictive resource (time!) constraints is better developed.

References

D.G. Bobrow, M. Stefik (1983)
The LOOPS Manual, XEROX Corporation

J. Breuker, B. Wielinga (1985)
KADS: Structured Knowledge Acquisition for Expert Systems, Proc. 5th Intern. Workshop on Expert Systems and Their Applications, Avignon, May 1985

J. Breuker, B. Wielinga
Use of Models in the Interpretation of Verbal Data, to appear in A. Kidd (ed.), Knowledge Elicitation for Expert Systems: a Practical Handbook, Plenum Press, New York

B.G. Buchanan, D. Barstow, R. Bechtal, J. Bennet, W. Clancey, C. Kulikowski, T. Mitchell, D.A. Waterman (1983)
Construction an Expert System, in F. Hayes-Roth, D.A. Waterman, D.B. Lenat, (eds.), Building Expert Systems, Addison-Wesley, Reading, Massachusetts, pp. 127-167

F. di Primio, D. Bungers, T. Christaller (1985)
BABYLON als Werkzeug zum Aufbau von Expertensystemen, in W. Brauer, B. Radig (eds.), Wissensbasierte Systeme - GI-Kongreß 85, Informatik Fachberichte 112, Springer-Verlag Heidelberg, pp. 70-79

M.S. Fox, S.F. Smith (1984)
ISIS: A Knowledge-Based System for Factory Scheduling, Expert Systems, vol. 1, no. 1

J.H. Griesmer, S.J. Hong, M. Karnaugh, J.K. Kastner, M.I. Schor, R.L. Ennis, D.A. Klein, K.R. Milliken, H.M. Van Woerkom (1984)
YES/MVS: A Continuous Real Time Expert System, Proc. AAAI-84

A. Hart (1985)
Knowledge Elicitation: Issues and Methods, Computer-Aided Design, vol. 17, 455-462

H. Haugeneder, E. Lehmann, P. Struß (1985)
Knowledge-Based Configuration of Operating Systems - Problems in Modeling the Domain Knowledge, in W. Brauer, B. Radig (eds.), Wissensbasierte Systeme - GI-Kongreß 85, Informatik Fachberichte 112, Springer-Verlag Heidelberg, pp. 121-134

F. Hayes-Roth, D.A. Waterman, D.B. Lenat, (eds.) (1983)
Building Expert Systems, Addison-Wesley, Reading, Massachusetts

C.F. Herot, G.P. Brown, R.T. Carling, M. Friedell, D. Kramlich, R.M. Baecker (1982)
An Integrated Environment for Program Visualisation, in H.-J. Schneider, A.I. Wasserman (eds.), Automated Tools for Information System Design, Proc. of the IFIP WG 8.1 Working Conf., New Orleans 1982, North-Holland Amsterdam, pp. 237-259

W.A. Martin, R.J. Fateman (1971)
The MACSYMA System, Proc. 2nd Symposium on Symbolic and Algebraic Manipulation, March 1971, pp. 59-75

J. McDermott (1982)(a)
XSEL: A Computer Salesperson's Assistant, in J.E. Hayes, D. Michie, Y.H. Pao (eds.), Machine Intelligence, vol. 10, pp. 325-337

J. McDermott (1982)(b)
R1: A Rule-Based Configurer of Computer Systems, Artificial Intelligence vol. 19, no. 1

H.-H. Nagel (1985)
Wissensgestützte Ansätze beim maschinellen Sehen: Helfen sie in der Praxis?, in W. Brauer, B. Radig (eds.), Wissensbasierte Systeme - GI-Kongreß 85, Informatik Fachberichte 112, Springer-Verlag Heidelberg, pp.

170-198

P. Raulefs (1984)
Knowledge Processing Expert Systems, in T. Bernold, G. Albers (eds.), Artificial Intelligence: Towards Practical Application, Proc. Technology Assessment Conf., Zürich, April 1984, North Holland Amsterdam, 1985, pp. 21-31

E.M. Riseman, A.R. Hanson (1985)
A Methodology for the Development of General Knowledge-Based Vision Systems, in W. Brauer, B. Radig (eds.), Wissensbasierte Systeme - GI-Kongreß 85, Informatik Fachberichte 112, Springer-Verlag Heidelberg, pp. 257-288

M.-J. Schachter-Radig (1985)
Wissenserwerb und -formalisierung für den kommerziellen Einsatz Wissensbasierter Systeme, in W. Brauer, B. Radig (eds.), Wissensbasierte Systeme - GI-Kongreß 85, Informatik Fachberichte 112, Springer-Verlag Heidelberg, pp. 314-332

M. Turner (1986)
Real Time Experts, Systems International, Jan. 1986, 55-57

W. van Melle, E.H. Shortliffe, B.G. Buchanan (1984)
EMYCIN: A Knowledge Engineer's Tool for Constructing Rule-Based Expert Systems, in B.G. Buchanan, E.H. Shortliffe (eds.), Rule-Based Expert Systems, Addison-Wesley, Reading, Massachusetts, pp. 302-328

W. Wahlster (1986)
The Role of Natural Language in Advanced Knowledge-Bases Systems, this book

D.A. Waterman (1985)
A Guide to Expert Systems, Addison Wesley, Reading, Massachusetts

The Role of Natural Language in Advanced Knowledge – Based Systems

Wolfgang Wahlster

Department of Computer Science
University of Saarbrücken
D – 6600 Saarbrücken 11
Federal Republic of Germany

Abstract: Natural language processing is a prerequisite for advanced knowledge-based systems since the ability to acquire, retrieve, exploit and present knowledge critically depends on natural language comprehension and production. Natural language concepts guide the interpretation of what we see, hear, read, or experience with other senses.
In the first part of the paper, we illustrate the needed capabilities of cooperative dialog systems with a detailed example: the interaction between a customer and a clerk at an information desk in a train station. It is shown, that natural language systems cannot just rely on knowledge about syntactical and semantical aspects of language but also have to exploit conceptual and inferential knowledge, and a user model. In the remainder, we analyze and evaluate three natural language systems which were introduced to the commercial market in 1985: Language Craft™ by Carnegie Group Inc., NLMenu by Texas Instruments Inc., and Q & A™ by Symantec Inc. The detailed examination of these systems shows their capabilities and limitations.
We conclude that the technology for limited natural language access systems is available now, but that in the forseeable future the capabilities of such systems in no way match human performance in face-to-face communication.

1. Introduction

Natural language processing is a prerequisite for advanced knowledge-based systems since the ability to acquire, retrieve, exploit, and present knowledge critically depends on natural language comprehension and production. Natural language concepts guide the interpretation of what we see, hear, read, or experience with other senses.
In artificial intelligence (AI) we are working under the assumption that natural language understanding and production are knowledge-based processes. Consequently, the existing natural language systems belong to the class of *knowledge – based AI systems,* which inter alia include expert systems and model-driven vision systems.

The preparation of this paper was supported by the SFB 314 Research Program of the German Science Foundation (DFG) on AI and Knowledge-Based Systems, and DEC's EERP. The contents of the paper benefited from discussions with Johannes Arz. Thanks go to Heide Dornseifer for typing and editing the manuscript and its revisions.

By definition natural language systems are no expert systems because every human knows at least one natural language without him having to be an expert for that reason.

The knowledge base of a natural language system includes both *linguistic* (e.g. lexicon, grammar, dialog rules) and *nonlinguistic* (e.g. a description of the objects in the domain of discourse) subparts. Whereas, ideally, the construction of the nonlinguistic part of the knowledge base is based on joint research of computer scientists and application specialists, the design and implementation of the linguistic parts relies on cooperation among computer scientists and linguists. For centuries linguists have gathered knowledge about various natural languages. In most cases unfortunately, this knowledge cannot be used directly in natural language systems because it is represented in computationally untractable formats or because it is not detailed enough for transformation into algorithmic systems. Thus, it is often a collaborative effort among linguists as 'experts for language' and computer scientists as 'experts for the formal representation of knowledge' to construct linguistic knowledge sources for natural language systems.

In AI a piece of software is called a *natural language system,* if

(a) a subset of the input and/or output of the system is coded in a natural language
(b) the processing of the input and/or the generation of the output is based on knowledge about syntactic, semantic, and/or pragmatic aspects of a natural language.

According to condition (a), there are natural language systems which combine e.g. pointing gestures on a terminal screen with linguistic descriptions. XTRA (eXpert TRAnslator, cf. Kobsa et al. 1986), a dialog system currently under development in our laboratory, will assist the user in filling in his annual income tax form. Here, a typical input contains a tactile gesture: *'Should I add my expenses for business trips here* ⟨pointing gesture on the income tax form displayed in a window of the terminal screen⟩ *or is it better to increase this* ⟨another pointing gesture⟩ *value?*.

Furthermore, most natural language systems commercially available today do not generate natural language. Their output consists of formatted data retrieved e.g. from a database or fixed text strings. Other systems use natural language input but generate graphical output. Yet another class of systems generates natural language descriptions for purely visual input.

Condition (b) is very important in that it rules out systems which process natural language purely as strings of characters without understanding its content (e.g. text editors, statistical packages).

It should be pointed out that today most natural language systems are based on written input and output. Although hardware for medium quality speech synthesis (without much prosodic features) is available, the speaker-independent recognition of continuous speech is still a major research problem.

2. Natural language understanding as a knowledge-based process

Let us use a simple example to demonstrate the role of knowledge-based access systems in advanced information systems (see Fig. 1).

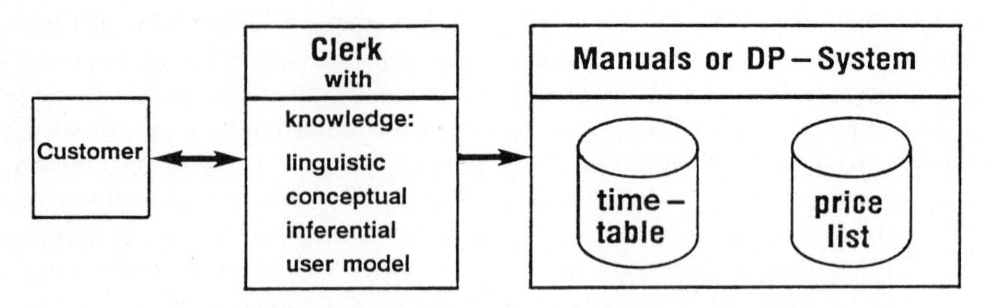

Usual situation at an information desk in a train station

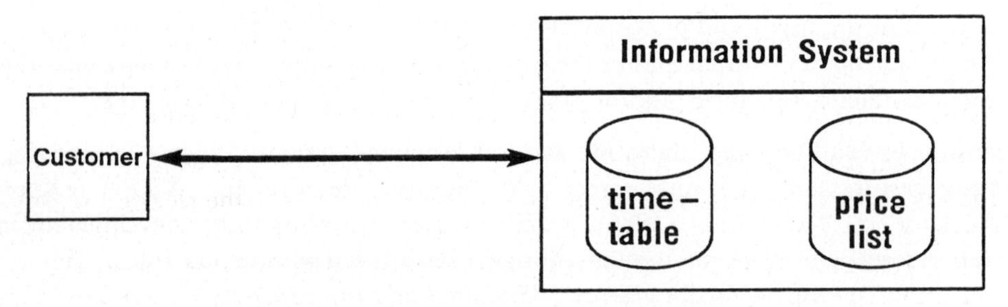

Unacceptable service quality as a result of the clerk's replacement by a conventional information system

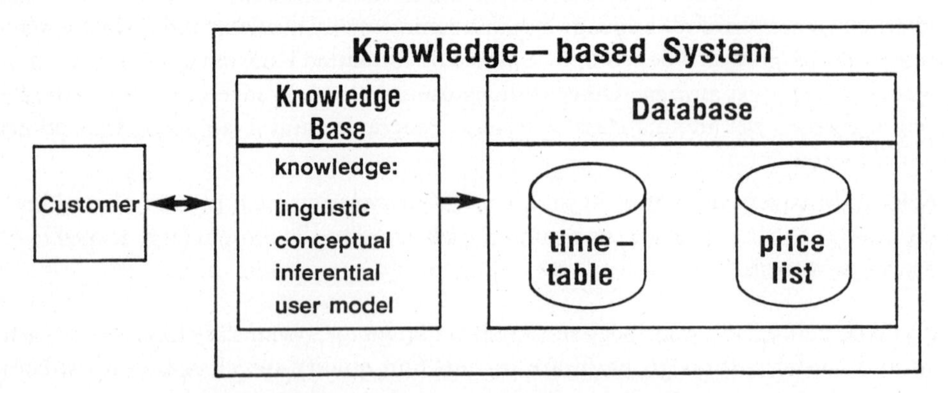

The combination of data – and knowledge bases results in an increased consultative capacity

Fig. 1: The need for knowledge-based consultation systems

Usually, a clerk at an information desk in a train station uses a time-table and a price list to respond to questions of an information-seeking customer. As formatted mass data, these tables and lists are contained in manuals or in a database system. As external data, they are not a part of his internalized knowledge. Since for the clerk, the access to these tables and lists is of critical importance, it is clear that he must be familiar with the organization of the manuals or the database.
For adequate consultation however, the clerk must, aside from such *access knowledge*, activate other areas of knowledge as well (see Fig. 1)

(a) If the customer says 'My son must be on the train to Saarbruecken on Monday. Does it have intercity connections?', he shortens his second sentence by using the pronoun 'it' instead of 'the train'. The clerk's *linguistic knowledge* helps him to select the correct referent for the pronoun. He is able to rule out 'My son', 'Saarbruecken', and 'Monday' as possible antecedents for 'it'. Furthermore, the clerk can use his linguistic knowledge about speech acts to recognize that the client does not just expect a yes/no answer, but also the departure times of suitable intercity trains to Saarbruecken.

(b) If the customer asks 'What is the difference between a sleeping-car and a couchette car' the clerk cannot find the answer in his time-table or the price list. But the clerk *can* use his *conceptual knowledge* to compare both concepts and to identify the distinguishing features of both alternatives for spending the night on a train.

(c) If the customer says 'I am going to Greece on an excursion together with my professor and our archeological seminar. Can I use a Eurorail ticket?', the clerk exploits his *inferential knowledge* referring to a rule like 'If the client is a student and is under 27 years old, then he can buy a Eurorail ticket'. He can apply this rule in a backward chaining mode, which means, that he has to test the if-part of the rule. Using an inference rule like 'If someone is attending a university course, then he is a student' the clerk can infer that the customer is a student. Then the clerk can respond with 'Yes' if the customer gives an affirmative answer to the clerk's question 'Are you under 27?'.

(d) If the customer says 'A return ticket to the Hanover Fair, please' the clerk will most probably offer him a first-class ticket. This response is based on a *user model* which the clerk derives from the assumption that a visitor to a professional fair is on a business trip. With the user class 'businessman' the clerk associates certain stereotypical knowledge, e.g. that travel costs are usually reimbursed for business trips. Thus the clerk assumes that a first-class ticket will be preferred by the customer.

In the 70's, there were many computer science projects which tried to replace the clerk by an information system. In such a scenario, the customer formulated his request in a query language and a DBMS evaluated the query and retrieved the relevant data from a database in which the time-table and the price list were stored (see Fig. 1) – the result of which being unacceptable service quality for most customers. For someone who uses a train twice a year, it is unreasonable to have to learn a formal query language (even if the query language is very simple like pushing a combination of four buttons out of a menu of 50) for getting information on railway connections. Even if the customer would spend the time to learn the query language, the lack of expressive power of current database query languages compared with natural language, e.g. the absence of indirect speech acts and anaphoric devices as exemplified in paragraph (a), makes such a dialog with an information system rather frustrating.

The questions presented in (b) - (d) above cannot be answered by a conventional information system based on database technology. No currently available DBMS can record and maintain complex conceptual knowledge structures, apply inference rules, and build up a user model as illustrated in examples (b) - (d).

From the AI point of view, the information system has to be complemented by a knowledge-based access system in order to become acceptable for the end-user (see Fig. 1).

In our example scenario, a knowledge-based consultation system could provide an increased consultative capacity by combining a database system with a knowledge base, containing a formal representation of the linguistic, conceptual, and inferential knowledge of the clerk, as well as stereotypes for user modeling.

Although presently for each problem mentioned in (a) - (d) at least one experimental knowledge-based access system which can adequately handle this type of question exists, it is, of course, a long way from implementing the broad-based universal communication capabilities of a clerk into an integrated and robust real time system.

A particular problem, far from being solved in any AI system, is the permanent *knowledge acquisition* of humans. In our example, one usually assumes that a clerk reads newspapers or watches TV news programs. Therefore he may know that this year the Davis Cup final takes place in Munich. Thus, if a customer asks for a return ticket to the Davis Cup final, the clerk will probably be able to offer him a return ticket to Munich. For a knowledge-based system today, it would be unrealistic to assume that the system daily updates its knowledge base with information potentially relevant for consultation purposes.

Although at present, the consultation capabilities of a knowledge-based system in no way match human performance, our example demonstrates that, if there were a real bottleneck in consultation capacity, a knowledge-based access system might help.

The example shows also that there are applications in which knowledge-based systems have to be combined with database systems. Without access to the time-table and price list, the best natural language system would be worthless in the scenario previously discussed. Furthermore, it would be clearly inadequate to represent the formatted mass data in one of the knowledge representation languages developed in AI.

In the following we analyze and evaluate three natural language systems which were introduced to the commercial market in 1985: Language Craft ™ by Carnegie Group Inc. (see Carnegie Group 1985b), NLMenu by Texas Instruments Inc. (see Texas Instruments 1985b) and Q & A ™ by Symantec Inc. (see Kamins 1985). It is interesting to note that each of these commercial systems was developed under the direction of a well-known scientist with considerable research experience in NL processing. The principle designer of Language Craft is J. Carbonell, who previously developed natural language systems such as XCALIBUR (cf. Carbonell et al. 1983) at Carnegie-Mellon University. NLMenu was designed by H. Tennant, who formerly was a member of the design team for the natural language system JETS (cf. Tennant 1980) at the University of Illinois at Urbana Champaign. G. Hendrix, as a principle designer of the natural language interfaces LIFER and LADDER (cf. Hendrix et al. 1978) at SRI International, has built the Q & A system for Symantec.

3. Language Craft: a transportable interface based on caseframes

Language Craft is a tool for constructing natural language interfaces to interactive systems, including databases and expert systems (see Carnegie Group 1985a, 1985b). Language Craft consists of PLUME™ and the Grammar Writer's Workbench™. PLUME includes a parser which contains some built-in knowledge of English morphology and syntax. For each new application, the developer of the interface has to write a domain-specific grammar, which mainly consists of a set of caseframes. The Grammar Writer's Workbench is a software tool for assistance in creating, editing, testing, and compiling domain-specific grammars.

In addition to a parser, which maps the natural language input onto a caseframe representation, PLUME provides a translator to convert the caseframes into Lisp function calls, driving the application. The rule-based translator works with two sets of rules, which the grammar writer has to specify. One set contains rules which collapse different parses into a single parse by rearranging caseframes to map into a canonical form. The other set of rules converts the caseframes into the actual Lisp forms to be passed on to the interactive application system.

Language Craft includes components for spelling correction and ellipsis resolution. For example, the system (S) uses the context provided by sentence (3) in Fig. 2 (see Carnegie Group 1985b) to fill in the missing information in (4).

When an input is ambiguous or a paraphrase is requested by the user (U), Language Craft uses a simple natural language generator which verbalizes the caseframes instantiated during parsing. As illustrated by example (6) in Fig. 2, it can paraphrase several readings of the input and ask the user to choose between them.

(1) U: Set up a Friday meeting for me at 2:00 with Mary Smith and Ron Brown.
S: Following meeting scheduled:
Friday March 29 2:00 to 3:00 Mary Smith, Ron Brown, Bob Jones

(2) U: What meetings does Smith have?
S: Monday March 25 10:00 to 12:30 J.P. Roberts, Mary Smith
Tues. March 26 4:00 to 5:00 Mary Smith, Bob Jones
Friday March 29 2:00 to 3:00 Mary Smith, Ron Brown, Bob Jones

(3) U: List all my meetings on Tuesday.
S: Tues. March 26 1:00 to 1:30 J.P. Roberts, Bob Jones, Jane Arran
Tues. March 26 4:00 to 5:00 Mary Smith, Bob Jones

(4) U: Friday meetings
S: Friday March 29 2:00 to 3:00 Mary Smith, Ron Brown, Bob Jones

(5) U: Cancel the meeting at 1pm Tuesday
S: Following meetings cancelled:
Tues. March 26 1:00 to 1:30 J.P. Roberts, Bob Jones, Jane Arran

(6) U: Give me 10:00 meeting with Smith Friday.
S: Sorry! Found 2 possible interpretations for your input:
1) Schedule a meeting for me with Smith on Friday at 10:00.
2) List a meeting for me with Smith on Friday at 10:00.
Please enter the number of the correct interpretation (0 for none of the above):
U: 1
S: Following meeting scheduled:
Friday March 29 10:00 to 11:00 Mary Smith, Bob Jones

Fig. 2: Sample dialog with a meeting scheduling system

The domain-specific grammatical knowledge is specified mainly in the form of case-frames. Caseframe definitions include syntactic knowledge which is expressed in a powerful pattern matching language. In addition to various pattern operators e.g. for disjunction, optionality, negation, assignment, and wildcards, the pattern language is comprised of so called non-terminal patterns.
Non-terminal patterns refer to complex patterns which are repeatedly used in a grammar.

For example:

(1) ⟨schedule⟩ → (schedule !! (&verb set) up !! arrange)

If ⟨schedule⟩ appears in the grammar, it is replaced by the right-hand side of the rewrite rule (1).
The parser treats '!!' as an 'exclusive or' and stops as soon as it finds a match with an element of the disjunction.

For regular cases, morphological processing is automatically done by PLUME, so that

in (1), the grammar writer only needs to specify the infinitive form for verbs. Thus, the terminal pattern 'schedule' in (1) will match e.g. 'schedules', 'scheduled' and 'scheduling'.
If a terminal pattern consists of several words (e.g. 'set up' in (1)), the grammar writer has to specify the word on which morphological transformations are to be performed (e.g. (&verb set) in (1)).

The rewrite rules for non-terminal patterns can be recursive, as illustrated by (2):

(2) ⟨directory⟩→ (%slash $?⟨directory⟩)

The operator '?' is used to indicate optionality and '$' is a wildcard which matches any single symbol. The symbol %slash represents the character '/' in a pattern. The above pattern matches UNIX™ directory names like '/man' or '/man/ray'.

Since the PLUME parser is mainly based on caseframe instantiation, it can handle inputs with odd word orders. PLUME distinguishes between sentential, nominal, and adjectival caseframes. Sentential caseframes usually represent sentences containing verbs, nominal caseframes represent noun phrases, and adjectival caseframes represent adjectives or relative clauses. Each caseframe contains a set of patterns (called header) which, if matched, activate the case frame. PLUME tries to fill the case slots of the activated case frames.
For every case, each caseframe definition contains a filler specification, which is either a pattern to look for in the input, or a name of another caseframe. There are other slots that tell the parser how to locate the case in the input, and in which order the cases should be checked.

The following example shows parts of the caseframes for 'schedule' and 'meeting':

```
[*schedule*
        :type sentential
        :header <schedule>
        :cases
            (initiator
                :positional subject
                :filler *name*)
            (meeting
                :positional direct - object
                :filler *meeting*)
            (day
                :type minor
                :semantic - class time
                :case - marker ?<on>
                :filler <days>)...]
```

```
[*meeting*
        :type nominal
        :header <meeting>
        :cases
            (participants - with
                :case - marker with
                :adjective *none*
                :filler *name*)
            (room
                :case - marker in
                :filler *room*)
            (duration
                :case - marker for !! of
                :filler *duration*)...]
```

The caseframe '*schedule*' refers to '*name*' and '*meeting*' as two nominal caseframes. It uses the non-terminal patterns ⟨schedule⟩, ⟨on⟩ and ⟨days⟩. The ':positional' slot of the initiator case tells the parser to identify the *right* most unmarked case to the *left* of the verb as a possible slot filler. The ':semantic-class' field, which has 'thing', 'person', 'place', and 'time' as possible values, helps the parser to handle Wh-questions with 'where' and 'when', and to differentiate between 'what' and 'who'. The ':case-marker' slot tells the parser that the pattern ⟨on ⟩ may immediately precede the filler ⟨days ⟩. Day is treated as a minor case, i.e. it is not examined unless there is input left over after the instantiation of all preferred and major cases. In the nominal caseframe '*meeting*' the value of the ':adjective' slot indicates that a prenominal use of this case is not permitted (e.g. 'the with Roberts meeting').

Given the input sentence 'Schedule a meeting with Schmidt in room 4' PLUME uses inter alia the caseframes shown above to construct the following caseframe instance:

```
{*schedule*
    (%mood declarative
        (meeting
            {*meeting*
                (%determination *indefinite)
                (participants - with
                    {*name*
                        (lname Schmidt)})
                (room
                    {*room*
                        (%number (singular))
                        (room - number 4)})
                (%number (singular))})
    (%voice active)
    (%tense (present))}
```

The parser scans the input and finds 'schedule' as a caseframe header of '*schedule*'. It picks up 'meeting' as a header of the direct object, filling the 'meeting' case in the '*schedule*' caseframe. In the '*meeting*' caseframe, it finds 'with' as a casemarker for the 'participants-with' case and 'in' for the 'room' case. In the '*name*' and '*room*' caseframes, it assigns 'Schmidt' and '4' as fillers to their slots 'lname' and 'room-number', respectively.

PLUME covers modes, tenses, voice, and negation automatically, so that the grammar writer only has to provide the caseframe definitions for each application domain.

The parser handles 'Yes/No' questions and 'who'-, 'what'-, 'where'-, 'when'-, and 'which'-questions using the same caseframes applied to parse declarative sentences.

Quantification and determination which are also covered by the grammatical knowledge built into the PLUME parser, appear as cases named '%quantification' and '%determination'. The quantifiers and determiners recognized by PLUME are 'most, almost all, more, any, each, none, few, all, every' and 'a, an, the, this, that, these, those, either, neither', respectively.

Release 2.0 of Language Craft does not include a pronoun resolution mechanism. It is implemented in Common Lisp and runs on Symbolics^{TM} Lisp machines and the VAX^{TM} under VMS^{TM}.

4. NLMenu: a Menu-based natural language interface

Presently, a major problem with most natural language systems is that they allow the user to input well-formed expressions which the system cannot understand. Thus, there is a mismatch between the user's linguistic competence and the system's limited coverage of natural language.
The basic idea of Menu-based NL interfaces like NLMenu is that instead of typing natural language in an unconstrained way, the user is guided by a set of dynamically generated menus to use the subset of natural language that the system understands (see Thompson 1984). The items of the menus are words, phrases, or so-called 'interaction experts', which are included in angle brackets (see Fig. 3).

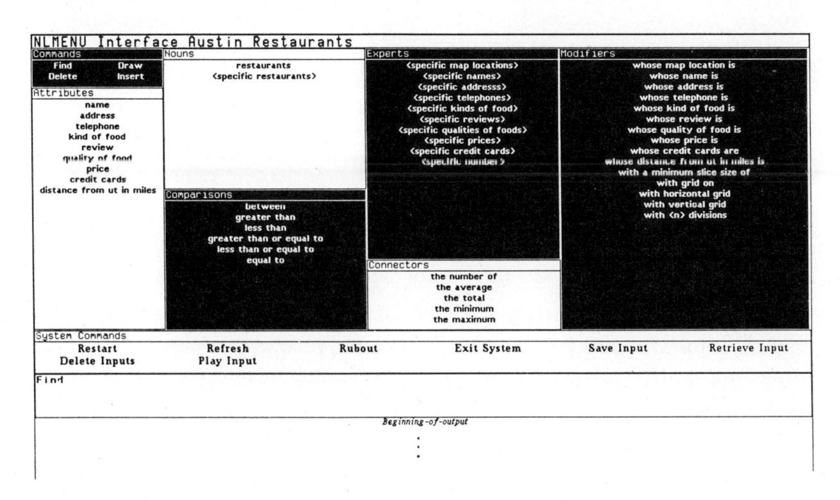

Fig. 3: The first selection phase

To select a menu item, the user moves the mouse cursor to the desired item and clicks on the mouse. After each selection by the user, NLMenu examines the natural language expression constructed so far, and decides which windows to activate next and which items to place in the windows. A selection of windows displayed in reverse video is not possible. As soon as the user has constructed a string, which presents a complete sentence according to the grammar controlling the NLMenu driver, the 'Execute' command appears as a selectable item in the window, displaying applicable commands.
This approach has the advantage that the user's input is generated to be both syntactically and semantically meaningful to the system. Figs. 3 - 5 (taken from Texas Instruments 1985a) illustrate the Menu-based approach using NLMenu as an interface to a restaurant database.
In Fig. 3, the user has selected 'Find' from the 'Command' menu; the only active window at the beginning of a new input. 'Find' appears in the sentence construction window at the bottom of the NLMenu screen shown in Fig. 3. NLMenu makes the 'Command' menu 'inactive' and activates the 'Attributes', 'Nouns', and 'Connectors' windows.

After two selection phases in which the user first chooses 'restaurants' from the 'Nouns' menu, and then, 'whose kind of food is' from the 'Modifiers' menu, NLMenu presents the screen shown in Fig. 4.

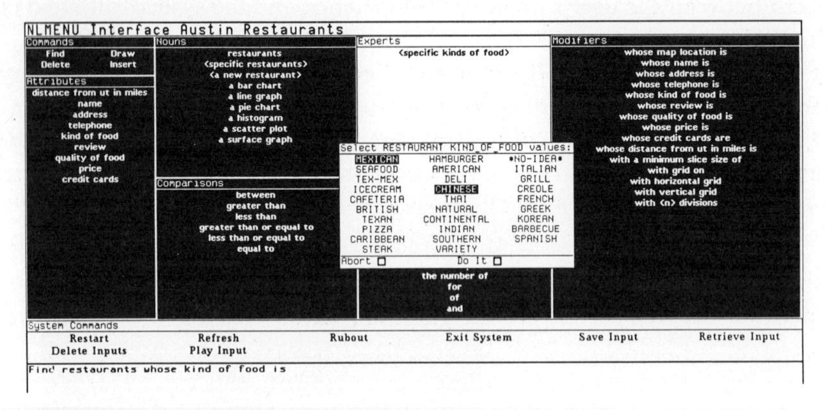

Fig. 4: Using an 'interaction expert'

If the user chooses '⟨specific kinds of food ⟩' a special window pops up with actual database values as shown in Fig.4. The user selects 'Mexican' and 'Chinese', so that his query now reads 'Find restaurants whose kind of food is Mexican or Chinese' (cf. sentence construction window in Fig. 5). NLMenu recognizes this as a complete sentence and presents the 'Execute' option in the 'Systems commands' window. The user may 'Execute' the query or continue it by selecting an item from the 'Connectors' menu. If he decides to 'Execute' his query, the system retrieves an answer from the database.

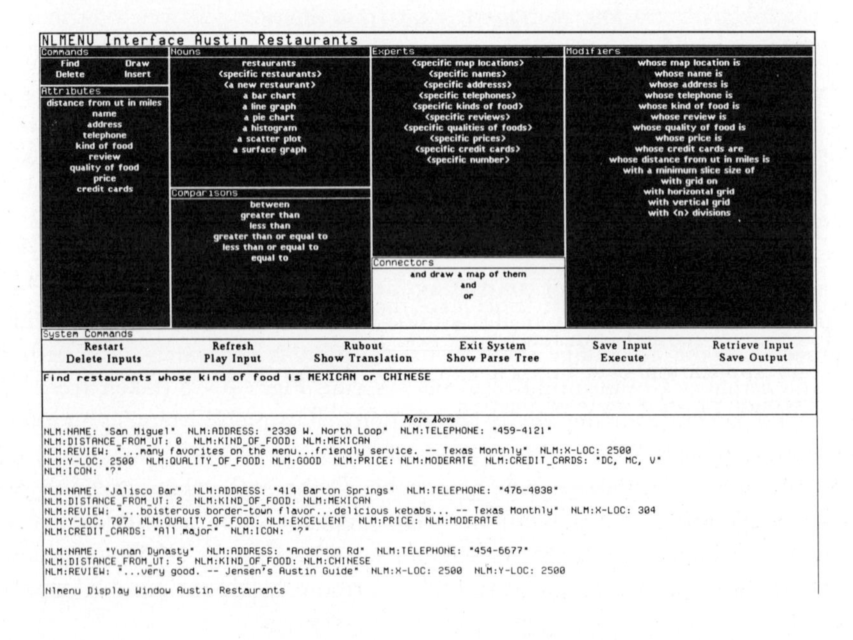

Fig. 5: The user executes the query

A common problem in natural language interfaces to databases is the recognition of database values when they appear in the user's input. Due to storage limitations, storing database values in the lexicon is not feasible for large domains. For example, consider an aircraft maintenance system for which the name of each aircraft has to be stored both in the database and the lexicon.
The developer of a NLMenu-based interface can use so-called interaction experts to support the user in referring to a specific database value. The 'interaction experts' are procedural augmentations to a context-free grammar. NLMenu provides a library of procedural 'interaction experts', e.g. for type-in windows or menu windows, in which items coming from the database are displayed. In Fig. 5, an 'interaction expert' was used to let the user specify a disjunction of a single-valued database field (Kind_of_food).

The most important knowledge sources, which the developer of a NLMenu interface to a new application has to specify, are the grammar and the lexicon. The grammatical knowledge is represented by an attributed context-free semantic grammar, which does not contain cycles and empty productions.
The basic grammar rule format is:

("$A \rightarrow B_1\ B_2 \ldots B_n$" (⟨Translation function 1⟩... ⟨Translation function m⟩))

The translation functions, which represent the semantic component of the grammar, are Lisp expressions where numbers correspond to the position of the symbols on the right-hand side of the rule.
The lexicon of NLMenu defines the items listed in the selection menus and relates them to the terminal symbols in the grammar. In addition, it contains information on the application-dependent meaning of a menu item in the form of so-called translations.

The basic lexical entry format is:

(⟨Terminal Symbol⟩ "⟨menu-item⟩" ⟨menu-window⟩ "⟨Translation⟩")

⟨Menu-window⟩ is the name of the menu in the selection window to which the item maps. ⟨Translation⟩ is a lambda expression or an atomic expression.

The translation functions specified in the grammar rules refer to the translations stored in the lexicon. For example, if a translation function reads (1 2 5) and the first, second and fifth symbol of the right-hand side of the corresponding rule is a terminal symbol of the grammar, the translation portion of the lexical entry for the first symbol is applied with the translation specified in the lexicon for the second symbol. That result becomes the function that is applied with the translation for the fifth element as its argument. Note that the application of a lambda expression to an argument produces another lambda expression or an atomic expression.

Consider a relational database which includes a relation RUNWAY with attributes 'Name' and 'Length'. Suppose we want to write a NLMenu interface to the database which is accessible via the SQL query language. Then, the task of NLMenu is to transform a natural language input like (1) into the SQL query (2).

(1) Find name and length of runways

(2) (SELECT NAME, LENGTH FROM RUNWAY)

In the following, this simple example is used to discuss the main idea behind the translation process.
Assume the developer of the NLMenu interface has specified the following somewhat simplified grammar (the parentheses indicate an optional sequence of symbols) and lexicon for this application:

Grammar:

(a) ("S → find MODIFIERS for MODIFIABLE-NP" ((4 (1 2))))
(b) ("MODIFIERS → ATTRIBUTE (attr-and MODIFIERS)" ((1) (2 1 3)))
(c) ("MODIFIABLE-NP → runways"((1)))
(d) ("ATTRIBUTE → name"((1)))
(e) ("ATTRIBUTE → length"((1)))

Lexicon:

(find "find" COMMAND "lambda f Select f from")
(name "name" ATTRIBUTES "name")
(length "length" ATTRIBUTES "length")
(attr-and "and" CONNECTORS "lambda a lambda b a,b")
(for "of" CONNECTORS nil)
(runways "runways" NOUNS "lambda r (r RUNWAY)")

Beginning with the lowest, rightmost subtree, the translator traverses the following parse tree built by the parser:

```
S (4 (1 2))

  (1) find
        "find"                      lambda f   Select f from

  (2) MODIFIERS (2 1 3)
      (1) ATTRIBUTE
            name (1)
               "name"
      (2) attr-and
            "and"                   lambda a lambda b   a,b
      (3) MODIFIERS
            ATTRIBUTE (1)
               length (1)
                  "length"

  (3) for
        "of"

  (4) MODIFIABLE-NP
        runways (1)
           "runways"                lambda r (r RUNWAY)
```

The translator then gets the translation list for the rule that constructed this subtree and gets the translations for the corresponding lexical entries. In our example it begins with rule (e). The corresponding translation list of lenght 1 has an atomic translation that is simply passed up the tree unchanged. The translator then moves up the tree one rule. After another translation step in which this atomic translation is passed up the tree (using rule (b) without the optional part), the translator applies the translation function (2 1 3) to the lexical entries for 'attr-and', 'name', and to the result of the previous translation which, in our example, is the atomic expression 'length'. The result of ("lambda a lambda b a,b" "name" "length") is "name, length". The translation function of rule (a) tells the translator to first apply the lambda expression representing "find" to the expression representing the translation of the MODIFIERS-subtree. This produces the result "Select name, length from". This expression is then used as an argument for the lambda expression specified in the lexicon as a translation for "runways". The application of "lambda r (r RUNWAY)" to "Select name, length from" produces the SQL query (2) as the final output.

NLMenu provides several software tools for the developer of a new interface (see Texas Instruments 1985b). The grammar writer's toolkit includes e.g. a conflict checker and a cycle checker for the grammar. Other tools assist in the completion and modification of the lexicon and in building the selection windows.

An interesting feature of NLMenu is the database interface generator which uses a NLMenu interface to acquire knowledge about a new database from the developer of a new application. When the developer has specified the necessary domain-dependent information, the interface generator uses a generic 'core' grammar and lexicon to automatically generate the complete grammar and lexicon for the application. The database interface generator is based on a two-level grammar approach. The 'core grammar' of NLMenu is a set of hyper-rules, i.e. rule templates that can be expanded to a set of context-free rules. The developer of a new interface only needs to specify values for a set of meta-rules, which in turn provide the slot fillers that instantiate the hyper-rules. Presently, the interface generator can only be used for simple relational databases since the 'core' grammar and lexicon of NLMenu is quite small.

A disadvantage of NLMenu is that the sentences constructed sound somewhat stilted. A principal reason for this is that offering various paraphrases would clutter the screen too much.

With NLMenu, no elliptical sentences can be directly constructed. But the user can point with the mouse to words or phrases in his previous input generated by interaction experts and change the items in the pop-up menus.

The current version of NLMenu does not provide morphological analysis, pronoun resolution or cooperative response generation. NLMenu runs on the Explorer^{TM} Symbolic Processing Systems produced by Texas Instruments.

5. Q & A: natural language database access on a personal computer

Q & A combines a text editor and a simple database system with a natural language interface in one software package. The natural language interface called 'intelligent assistant' (IA) supports retrieval and update operations on the database.
Unlike Language Craft and NLMenu which run on powerful, but relatively expensive workstations or mainframes, Q & A runs on IBM Personal Computers or compatibles requiring a minimum of 512k RAM.

In contrast to the general relational data model, databases in Q & A are restricted to a single relation. In Q & A, each tuple of the relation stored in the database is called a 'form'. An attribute of a relation is referred to as a 'field'. The field values of a Q & A database are taken from a fixed set of domains which includes 'Numbers', 'Money', 'Dates', 'Hours', 'Text strings', and 'Yes/No'. In the database schema, which degenerates to a form template in Q & A, the user has to specify the names of the fields together with the corresponding domains from which values can be taken to fill out fields of a form.

IA has a built-in vocabulary of some 400 words. In addition, it uses the database-as-dictionary approach, i.e. all nonnumeric field values of the stored forms are inverted, providing an index of all the words and phrases used in the database and the fields in which they occur. The information contained in the database schema (field names and associated domains) is also used as part of the lexicon. Finally, there is an interactive knowledge acquisition component based on a 'Lesson Menu' in which the user can specify:

- a set of generic terms that refer to forms in the database (e.g. {employee, staff member, person}, {enrollment, sign up})

- a set of unique alternate field names (e.g. {supervisor, boss} for the 'manager' field specified in the database schema)

- a set of verbs, which like alternate field names, refer uniquely to a field (e.g. the verb 'earn' can be associated with the 'salary' field, so that 'How much does John earn' is interpreted as a paraphrase of 'What is John's salary')

- measurement terms that can be associated with numeric fields (e.g. 'days' for the numeric field 'accrued vacation')

- a set of adjectives which can be associated with specific fields form the domains 'Numbers' and 'Money'. For each new adjective, the user has to specify whether it describes a low value or a high value on the according numerical scale. Note that this classification is field-specific, e.g. 'excellent' may refer to a low value of the 'gas consumption' field but a high value of the 'salary' field. The semantic interpretation for the unmarked forms of the adjectives is 'higher (or lower) than average', for comparative forms 'higher (or lower) than the comparative value', and for superlatives 'the maximal (or minimal) value'.

In the Menu-based knowledge acquisition dialog, the user can also provide IA with three types of semantic information not contained in the database schema by marking fields that

- can be used as definite descriptions, i.e. that uniquely identify a specific form in the database (these fields are included in all detailed reports)
- contain locations (required for answering 'Where is'-questions)
- contain people's names (e.g. first name, last name, title; required for answering 'Who'-questions).

The user of IA can define various sorts of synonyms and paraphrases, which he may introduce during the natural language dialog (e.g. by 'Define total pay as salary plus bonus) or in a stylized subdialog. Words like 'well', which may be ignored in the input, can be defined as 'empty' synonyms (e.g. 'Define well to be""').

Whereas the user of Q & A can apply the various methods discussed above to enhance IA's lexical knowledge, he cannot change or extend the grammatical knowledge of the system.

The built-in grammatical knowledge of IA is represented by an attributed context-free grammar which, in the distributed version, is compiled into a more efficient but equivalent form. IA works with a semantic grammar, i.e. most nonterminals are semantic categories specific to the database application. Left-recursion is not allowed in the rules. During a parse of a sentence like 'Add 100 $ to Schmidt's bonus' the following rule is applied:

```
⟨Revision⟩→ (ADD ⟨T4⟩ TO ⟨T4_2⟩)
             (SHOW_FIELD (COMPOSE_TERM ⟨T4⟩ '+ ⟨T4_2⟩))
```

The right-hand side of a rule can include individual words ('ADD' and 'TO' in the example above), predicates (e.g. for numbers) and/or nonterminals (⟨T4⟩and ⟨T4_2⟩in the example above).

Associated with each rule is a LISP expression, specifying a structure to be produced by the parser. Because the nonterminals are used as variables in the attached LISP code, it is necessary to introduce new terminals if otherwise there would be a conflict (as in the above example). The LISP expression can also contain conditions which, if not met, can have the effect of blocking a rule.

Before the parser is applied to the user's request, a scanner maps the input onto a canonical representation. The scanner includes a morphological component which recognizes regular verb endings, regular comparative and superlative forms of adjectives, and regular plurals of nouns and possessives. It substitutes the definition for all synonyms and paraphrases recognized in the input. Whereas all other morphological information is discarded by the procedure called 'lexical stripper', only the information contained in the suffices for comparatives, superlatives and possessives is forwarded to further processing stages.

The backtracking parser, which is implemented in IQLISP, transforms the natural language input into a sequence of so-called 'packets'. Packets are words or phrases in the input which form a semantic unit corresponding to constructs of Q & A's data manipulation language. For example, the input (1) is transformed into the packet sequence (2).

(1) What is the average age of females with salary >30000.
(2) ((AGG A) (? AGE) (R SEX =FEMALE) (R SALARY >30000))
(3) ((TYPE SUPPRESSED)
    (FIELDS (AGE A))
    (RESTRICTION (SEX =FEMALE) (SALARY >30000)))

Here, the parser has recognized four packets: the aggregate (AGG) function 'average' (A), the database field 'AGE' as the focus of the request, and two restrictions (R) on the field values.

A component called 'semtop' translates this sequence of packets into a semantic representation called 'logical form' (see (3) above). This representation is the input to a 'rephrase' procedure which generates a stylized English paraphrase of what the system intends to do. If the user gives his OK for the paraphrase, the logical form is input to a procedure which generates the so-called functional form of the user's request. The result of this final translation process is submitted to the database management system.

IA has no problems with partial sentences, i.e. input in a telegraphic style like 'Secretary's salaries' (which IA interpretes as 'Show me the names and salaries for all secretaries'), because its semantic grammar is not based on the notion of complete sentences but on packets as smaller units.

If a phrase in the user's input is ambiguous, IA highlights it and asks the user for clarification. It presents a numbered list of the paraphrases of all readings found, and the user can select the intended request by one of the numbers.

When encountering an unknown word which is not enclosed in quotes (used in update requests for marking new information), IA highlights it and asks the user for a clarification or a correction.

IA's built-in grammar is relatively large (more than 250 nonterminals) and covers a significant portion of possible retrieval and update operations on Q & A databases. It copes with declarative, imperative, and interrogative sentences. The system handles 'What'-, 'Who'-, 'Where'-, and 'How many' -questions. It covers restrictive relative clauses and some forms of conjunction and disjunction on the level of noun groups, verb groups, and clauses, but excludes embedded possessives like 'Gary's manager's age'.

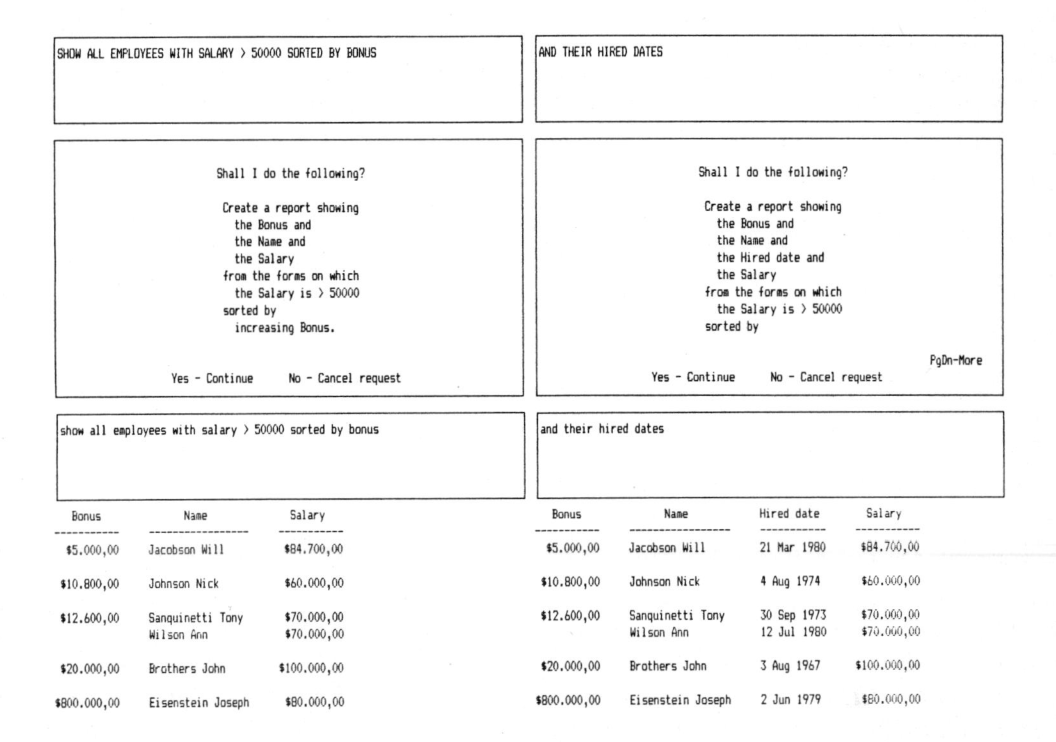

Fig. 6: A sample interaction with Q & A

IA can handle simple follow-up questions (see Fig. 6) which include singular or plural forms of personal pronouns (e.g. they) or possessive pronouns (e.g. her). Only the most recent request/response pair is checked for antecedents of a pronoun.

The instruction manual gives a crisp summary of IA's limitations: "You are infinitely smarter than the assistant, have a far larger vocabulary, and possess a much stronger grasp of the English language" (see Kamins 1985, p. IA-3).

6. Concluding remarks

Fig. 7 summarizes the distinguishing features of the three systems discussed in this paper. It is important to keep in mind that a 'Yes' in the table does not indicate, that the corresponding problem has been solved in its full generality.
The commercial systems compared in Fig. 7 offer the promise of easy transportability from one domain to another by providing various tools for easy acquisition of domain-specific information. They try to have the domain-dependent knowledge sources clearly seperated from the more general domain-independent knowledge sources.
Two of them (NLMenu and Language Craft) go beyond domain-independence in the attempt to provide a core system that is not restricted to database access but can serve as

	Language Craft	NLMenu	IA
Handling input with odd word order	Yes	Not possible	Yes
Handling partial input/ellipsis	Yes	No	Yes
Simple pronoun resolution	No	No	Yes
Generation of paraphrases	Yes, if input is ambiguous or on user's demand	No	Yes, always
Spelling corrector	Yes	Not necessary	No
Recognition of database values in the input	must be included in the lexicon	'interaction experts'	database is used as an extension of the lexicon
Morphological analysis	Yes	No	Yes
Expandable grammar	Yes	Yes	No
Type of grammar	patterns in caseframes	attributed context-free grammar	attributed context-free grammar
Type of parser	caseframe instantiation	bottom-up, incremental	top-down, backtracking
Target language of parser	instantiated caseframes	application-dependent	logical form
Target system	interactive software	interactive software	Q&A databases

Fig. 7: A functional comparison of three commercially available NL interfaces

a basis for interfaces to a variety of interactive systems (e.g. expert systems, command languages).
One of the major prerequisites for a wide-spread application of most language access systems is that the system allows its adaptation to new applications by nontechnical people who know the application domain, but are neither experts on the system nor specialists in AI or computational linguistics. This prerequisite is not yet fulfilled by the transportable systems NLMenu and Language Craft. On the other hand, Q & A fulfills this criterion, but it is restricted to database access. Moreover, some knowledge about a particular database system is built in at a very low level of processing.

A general-purpose natural language dialog system should be adaptable to applications that differ not only with respect to the domain of discourse, but also to dialog type, user type, and intended system behavior. In Wahlster and Kobsa 1986, we call such systems, which are transportable and adaptable to diverse conversational settings, *transmutable systems*. A first attempt to build a transmutable system was our design of the experimental dialog system HAM-ANS (see Hoeppner et al. 1984), whose dialog behavior can be switched from a 'cooperative' mode (e.g. the system answers questions about a traffic scene) to a 'interest-based' mode (e.g. the system tries to persuade the user to book a room in a particular hotel).

When people communicate, they do so for a purpose specific to the conversational situation. On the other hand, the systems discussed in this paper have no interest beyond providing the information-seeking user with relevant data. In the long run, natural language systems as components of advanced knowledge-based systems must perform a greater variety of illocutionary and perlocutionary acts: they may teach, consult, or persuade the user, inspire him to action or argue with him (see Bates and Bobrow 1984, Wahlster 1984, Webber 1986, Woods 1984). The major problem builders of transmutable systems are confronted with, is the lack of a representational vocabulary for the declarative description of the relationship between the system and the user, the system's intended dialog behavior and the associated conversational tactics.

Although in the forseeable future, natural language AI systems will not be able to behave exactly like the clerk at the information desk, discussed in the first part of this paper, the technology for transportable natural language access systems is available now and begins to have a major impact on the design and application of man-machine systems.

References

Bates, M., and R.J. Bobrow (1984): Natural Language Interfaces: What's Here, What's Coming, and Who Needs it. In: Reitman, W. (ed.): Artificial Intelligence Applications for Business. Norwood: Ablex, 179-194.

Carbonell, J.G., W.M. Boggs, M.L. Mauldin, P.G. Anick (1983): The XCALIBUR Project: A Natural Language Interface to Expert Systems. In: Proceedings of the International Joint Conference on Artificial Intelligence, Karlsruhe, West Germany, 653-656.

Carnegie Group (1985a): Language Craft^{TM}: An Integrated Environment for Constructing Natural Language Interfaces. Pittsburgh, PA.

Carnegie Group (1985b): The Language Craft^{TM} Manual. Release 2.0. Pittsburgh, PA.

Hendrix, G.G., E.D. Sacerdoti, D. Sagalowicz and J. Slocum (1978): Developing a Natural Language Interface to Complex Data. In: ACM Transactions on Database Systems.

Hoeppner, W., Th. Christaller, H. Marburger, K. Morik, B. Nebel, M. O'Leary and W. Wahlster (1983): Beyond Domain-Independence: Experience with the Development of a German Language Access System to Highly Diverse Background Systems. In: Proceedings of the International Joint Conference on Artificial Intelligence, Karlsruhe, West Germany, 588-594.

Kamins, S. (1985): Instruction Manual Q & A. Version 1.0. Document No.: 1-001. Cupertino: Symantec.

Kobsa, A., J. Allgayer, C. Reddig, N. Reithinger, D. Schmauks, K. Harbusch, W. Wahlster (1986): Combining Deictic Gestures and Natural Language for Referent Identification. Technical Report, SFB 314, Dept. of Computer Science, University of Saarbruecken, West Germany (forthcoming).

Tennant, H. (1980): Syntactic Analysis in Jets. Advanced Automation Group Coordinated Science Laboratory, University of Illinois at Urbana-Champaign. Working Paper 26.

Texas Instruments (1985a): Explorer^{TM} Natural Language Menu System. Data Systems Group, Technical Report No. 2533593-0001, Austin, Texas.

Texas Instruments (1985b): Natural Language Menu User's Guide. Digital Systems Group, Dallas, Texas.

Thompson, C.W. (1984): Using Menu-Based Natural Language Understanding to Avoid Problems Associated with Traditional Natural Language Interfaces to Databases. Texas Instruments, Computer Science Laboratory, Technical Report No. 84-12, Dallas, Texas.

Wahlster, W. (1984): Cooperative Access Systems. In: Future Generations Computer Systems, Vol.1, No.2, 103-111.

Wahlster, W. and A. Kobsa (1986): Dialog-Based User Models. In: G. Ferrari (ed.): Special issue on natural language processing. IEEE Proceedings.

Webber, B.L. (1986): Questions, Answers and Responses: Interacting with Knowledge Base Systems. In: Brodie, M., J. Mylopoulos (eds.): On Knowledge Base Management Systems. New York: Springer.

Woods, W.A. (1984): Natural Language Communication with Machines: An Ongoing Goal. In: Reitman, W. (ed.): Artificial Intelligence Applications for Business. Norwood: Ablex, 195-209.

Cognitive Science:

Information Processing in Humans and Computers

Gerhard Fischer

Department of Computer Science
and
Institute of Cognitive Science
University of Colorado, Boulder

Abstract

In the world of today the construction of complex systems cannot be limited to what is technologically feasible but has to take into account what is desirable and manageable by humans. Theories and methodologies from Cognitive Science (being an interdisciplinary research area whose goal is to create a better understanding of a broad range of cognitive abilities like design, understanding, learning, remembering and thinking) are an important source of knowledge and insight towards the construction of user-centered systems.

In the last ten years we have designed and constructed a variety of systems and tools in the areas of knowledge-based systems, human-computer communication and intelligent user support systems. Human factors considerations and results from Cognitive Science have served as design principles. Simultaneously the construction of these systems has contributed towards further progress in Cognitive Science by generating a large number of interesting and challenging problems for it.

1. Introduction

The microelectronics revolution of the 1970s made computer systems cheaper and more compact, with a greatly increased range of capabilities. Computing moved into the workplace, into schools and into homes for the potential benefit of everyone. Much of the available computing power is wasted, however, if users have difficulties in understanding and using the full potential of these systems. In the past too much attention has been given to technical aspects which have provided inadequate solutions to real world problems, imposed unnecessary constraints on users and been to rigid to respond to changing needs. More 'intelligent' software is needed which has knowledge about the user, the tasks being carried out and the nature of the communication process.

The human should be the "fixpoint" for the design of human-computer systems. Insight and knowledge from Cognitive Science should be used to construct user-centered systems which are best understood as embedded systems. In this article we will characterize the interrelationships between research in Cognitive Science, Artificial Intelligence and Human-Computer Communication. Specific examples to illustrate some of the general hypotheses, claims and ideas are drawn mostly from our research.

2. Human-Computer Systems - How they are and how they should be

2.1 Human-Computer Systems - How they are

Many aspects of human-computer systems have not kept pace with the dramatic progress in the area of hardware: especially how easy it is - not only for the expert but also for the novice and the occasional user - to take advantage of the available computational power *to use the computer for a purpose chosen by him/herself.*

Most computer users feel that computer systems are unfriendly, not cooperative and that it takes too much time and too much effort to get something done. They feel that they are dependent on specialists, they notice that "software is not soft" (i.e., the behavior of a system can not be changed without a major reprogramming of it), and the casual users find themselves in a situation similar to instrument flying: they need lessons (relearning) after not using the system for a while.

Our empirical investigations have shown that most users use only a small percentage of the functionality of current computer systems. To increase the computational power at the hardware level in future systems is a meaningful goal only if we can find and develop new ways to increase the usability factor of a system.

2.2 Human-Computer Systems - How they should be

The major challenge of designing and introducing more powerful computer systems is not primarily a technical one. *User-centered system design* [Fischer 84a; Norman, Draper 85] starts on the 'outside' by taking into consideration the overall social and technical environment, by examing human psychological and behavioral needs and then moves inwards to the specific technical details (see Figure 2-1).

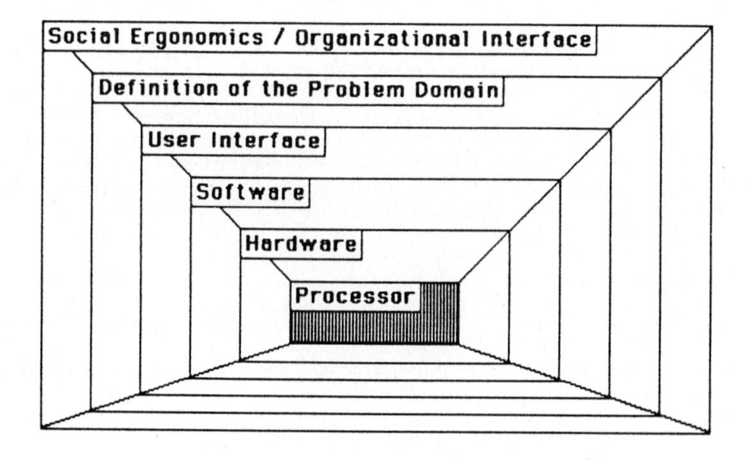

Figure 2-1: System Development from the Outside to the Inside

In the past, methodologies for creating and assessing computer systems were *computer-centered*, which is why so many failed to match their actual operating environments to user needs.

Our goals are to create human-computer systems which

1. handle all of their user's information-related needs; these needs will be quite different for different groups of users,
2. make computer systems accessible to more people and make computer systems do more things for people,

3. help us to gain a better understanding of cognitive dimensions and tasks structures.

To achieve these goals we need beside technological expertise

1. theories, which take knowledge and insights from Cognitive Science into account and help us to define new design criteria,
2. methods which are based on those theories,
3. tools which use the computational power of the computer to support the user (see Figure 5-2).

3. Cognitive Science

3.1 The Development of Cognitive Science

Cognitive Science is a group of related disciplines (bringing together researchers from Computer Science, Artificial Intelligence, Psychology, Linguistics, Sociology and Philosophy) characterized by a common approach to complex information-processing systems. Simon [Simon 81a] defines Cognitive Science as *"the domain of inquiry that seeks to understand intelligent systems and the nature of intelligence"*. It emphasizes detailed investigations and explicit theoretical models of mechanisms underlying complex symbolic tasks. In many ways, Cognitive Science is not really a new discipline: it deals with phenomena of thought and language that have occupied philosophers and scientists for thousands of years. During the last ten years several organizations, journals, institutes etc. have been founded and formed to give the field more structure and more visibility.

Its character can best be understood by comparing it to some closely related fields:

1. *Cognitive Psychology* (being a substantial part of Cognitive Science) is a branch of traditional experimental psychology which follows specific methodological principles that limit its scope.

2. *Information Processing Psychology* is formed on the assumptions that can best be understood by analogy to computers. It relies on the principle that cognitive systems are symbol systems which share a basic underlying set of symbol manipulation processes and that a theory of cognition can be couched as a program [Newell, Simon 76].

3. *Artificial Intelligence* designs and tests complex computer programs that ex-

tend the capabilities of computers. It is primarily a normative science, which is not only interested in understanding intelligence but improving it.

4. *Cognitive Simulation* moves towards an understanding of human intelligence by constructing and analyzing process models of human behavior.

3.2 Information Processing in Humans

Norman [Norman 81] illustrates "how little is known about so much of cognition". We lack a detailed and profound understanding of many cognitive abilities like perception, memory, learning, problem solving, performance and skilled behavior. Despite the lack of precise models, a careful analysis of human behavior should play a major role in the design of future human-computer systems. To improve human-computer communication, a detailed analysis of successful human communication processes can provide important guidelines. Human conversational partners

- share a lot of information (this common pre-understanding lets us communicate with a minimum of words and conscious effort),
- can model one another's knowledge and capabilities,
- can process huge amounts of information (even conflicting information),
- can update all of these structures (as the conversation progresses).

The understanding of human information processing has led us to the following qualitative design criteria for human-computer systems:

1. *The limiting resource in human processing of information is human attention and comprehension, not the quantity of information available.* Modern information and communication technologies have dramatically increased the amount of information available to individuals. Two important aspects of human-computer communication are the ability to access important information and to choose its appropriate representation.

   To illustrate this claim, we can use an example from modern aircraft design (see [Chambers, Nagel 85]): there are 455 separate warnings on a Boeing 747. Without display systems which consists not only of instruments, but which have the ability to prioritize information before presenting it to the crew, there is little hope of avoiding an information overload for the pilot.

2. *In complex situations, the search for an optimal "rational" solution is a waste of time.* There are limits to the extent to which people can apply rational analyses and judgements in solving complex, unpredictable problems. It is insufficient to ask people to *"think more clearly"* without providing new tools, such as knowledge-based systems and better human computer

communication, which have the potential to extend the boundaries of human rationality. The aim is to achieve the most satisfactory solutions given current knowledge, accepting that "better" solutions will emerge as the result of experience and enhanced knowledge and understanding.

User interface management systems (see Figure 3-1), programming environments and tools to deal with the construction and utilization of knowledge-based systems (like knowledge editors and browsers (see Figure 6-3)) are indispensable for coping with complex information processing systems.

3. *The limitations and structure of human memory must be taken into account in designing human-computer communication.* People have limited short-term memories. The way people recognize information is different from how they recall memory structures. This distinction, for example, is relevant to judging the advantages and limitations of different interaction models, such as comparing a command-based interface to a menu-based interface (see Figure 3-1).

   Our intelligence has become partially externalized, contained in artifacts as much as in our head: the computer is, in some sense, an artificial extension of our intellect, invented by humans to extend human thought processes and memory.

4. *The efficient visual processing capabilities of people must be utilized fully.* Traditional displays used with screen-based workstations have been one-dimensional, with a single frame on the screen usually filled with lines of text. New technologies have opened ways of exploiting human visual perception through the use of multi-window displays, color, graphics and icons.

   Figures 6-1 and 6-2 show some of our visualization tools (which are discussed briefly in section 6.1) which illustrate the anatomy of complex LISP data structures.

5. *The structure of the computer system must be understandable by people using it rather than requiring the user to learn by rote the functions that can be performed.* An adequate understanding of how a system works gives users the knowledge and confidence to explore the full potential of a system, which can have a vast range of possible options. Learning by rote may train the user to operate a limited number of functions but makes it difficult for the user to cope with unexpected occurrences and inhibits their exploitation of the full potential of the system.

6. *There is no such thing as "the" user of a system: there are many different kinds of users and the requirements of an individual user grow with experience.* Computer systems built based on a static model of the user are too rigid and limited to meet the demands of a diverse user community or the evolving needs of each user. In [Fischer 84b] we discuss system architec-

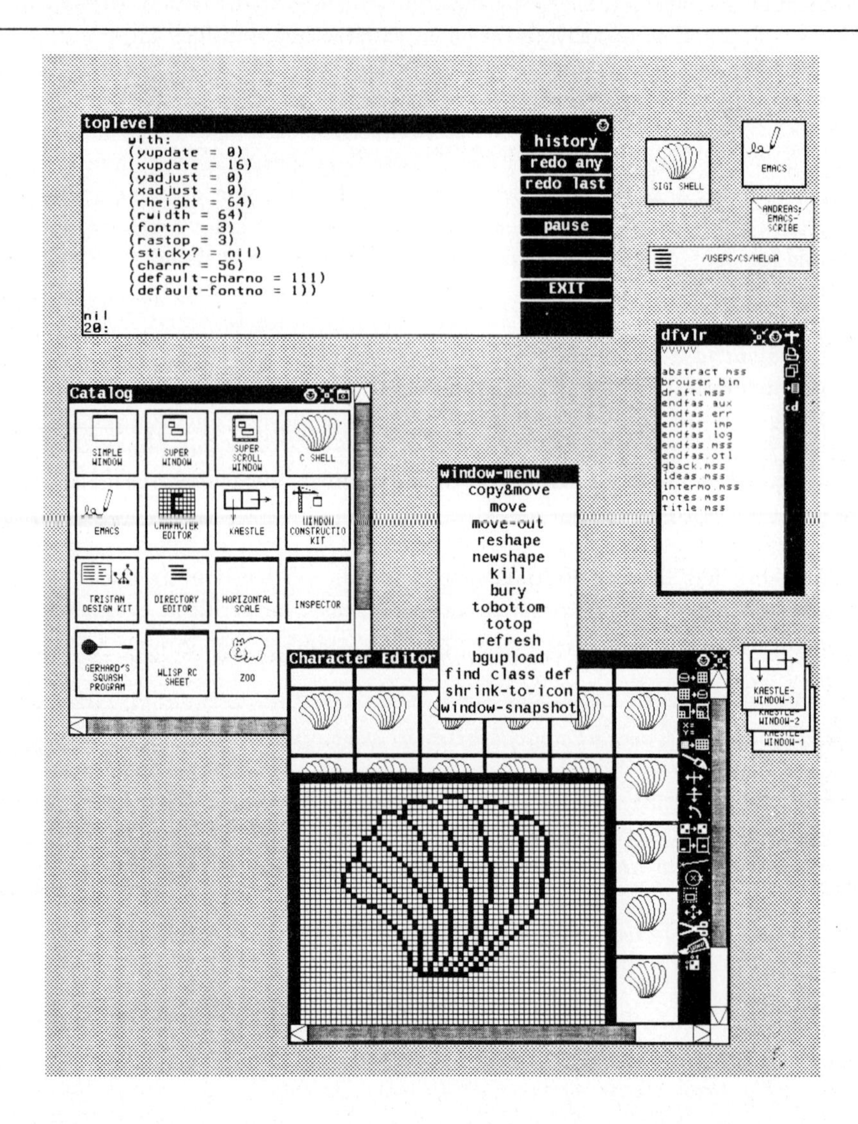

Figure 3-1: A Screen Image of our User Interface Management System

A LISP interpreter runs in the **toplevel** window; different systems can be activated by selecting the appropriate icon from the **Catalog**; new icons or characters can be defined with the **Character Editor**; a **window-menu** is associated with each window; windows can be shrunk to icons when they are not needed (some of them are shown on the right side). The **directory-editor** window gives access to the UNIX file system. The interaction style is based on manipulating the "world" (represented on the screen) *directly* by using a mouse.

tures which are able to provide some adaptivity and in [Fischer, Lemke, Schwab 85] we describe a prototypical system which treats users as individuals.

3.3 Information Processing in Computers

The original view of the computer is given by its name: *a device to do computation.* In the mid fifties the original workers in Artificial Intelligence showed that this view was much too limited: the computer was not only a giant adding machine but a *general symbol manipulation device.* They started to develop heuristic programs (in contrast to algorithms) to show that computers could be used to solve problems where it was not known exactly how to solve them. A major outcome from this shift in perspective was the invention and routinization of manipulable symbolic representations (starting with IPL-V, pushed forward tremendously by LISP and being extended currently by formalisms for knowledge representation).

Another view of information processing in computers is that they can be of great help to humans to assist them in many activities. Higher level programming languages, interactive programming environments, editors, document production systems and message systems are general information processing tasks which require little computation in the traditional sense. The computer serves as a *structured dynamic media for communication* which is qualitatively different from earlier media like pencil and paper or telephone.

Seeing the computer as a general symbol manipulation device has created concepts which have proven useful for disciplines which deal with human information processing. Computer-based languages provide the best known formalisms to describe information processing in humans and they allow us to formulate more precise models about human cognitive functioning.

4. A Cognitive Science Perspective of some Current Research Efforts in AI and Human-Computer Systems

Research in Cognitive Science should not only help us to understand our own human nature, but it should also help us in our practical affairs. It should provide the scientific understanding for a cognitively oriented approach towards the design of human-computer systems, sometimes called "Cognitive Engineering". The epistemological adequacy of a formalism is primarily not a formal or theoretical issue (theoretically almost all formalisms and programming languages are Turing computable) but a cognitive issue. For many problems, the question of *subjective computability* (to find ways to create new systems for complex tasks which we were unable to tackle in the past) is more relevant than whether a problem is computable in theory. Criteria to differentiate and evaluate different frameworks for knowledge representation (e.g. rule-based, object-oriented, logic-based and hybrid systems) have to be derived not primarily from formal language theory but from Cognitive Science and Task Analysis.

Following, we will briefly discuss natural language interfaces, expert systems and cognitive ergonomics research from a Cognitive Science perspective.

4.1 Natural Language Interfaces

Responsible proponents of natural language interaction acknowledge that current programs cannot understand language in a significant sense [Winograd, Flores 86]. This does not rule out the use of natural language interfaces because many practical applications (e.g. access to databases) do not demand deep understanding. The practicality of limited natural language systems is still an open question. Since the nature of the queries is limited by the formal structure of the data base, it may well be more efficient for a person to learn a specialized formal language designed for that purpose, rather than learning through experience just which natural language sentences are and are not accepted. The *habitability* [Bates, Bobrow 84] of a system (which measures how quickly and comfortably a user can recognize and adapt to the system's limitation) is a critical issue which needs to be studied empirically in realistic situations.

One may argue that there exists currently a *front-end fallacy*. Many human-computer systems have to perform more sophisticated functions than answering requests for factual information (e.g. helping users to formulate their problems, assisting in cooperative problem solving). These tasks require more elaborated data models and knowledge representations [Williams et al. 82] and additional types of reasoning. An appropriate interactive behavior will not be realized merely by tacking some off-the-shelf front end onto an existing system.

Another objective in natural language interface research is not so much to be able to analyze ever more complex sentences involving increasingly difficult semantic concepts, but rather to understand the process of intention communication and recognition well enough to enable a system to participate in a natural dialogue with its user [Winograd, Flores 86]. Assuming we had a natural language interface to UNIX [Wilensky et al. 84], we probably would be unpleasantly surprised if our question *"How can I get more disc space?"* would be answered by *"Type rm *"* - despite the fact that this would solve our problem (probably not in the way we intented; 'rm *' deletes all files in a directory). The problem in human-computer interaction is not simply that communicative trouble arises that does not occur in human communication, but rather that when the inevitable troubles do arise, there are not the same resources available for their detection and repair.

Based on the fact that there exists an *asymmetry* between human and computer (see Figure 5-1), the design of the interface is not only a problem of simulating human to human communication but of engineering alternatives to interaction related properties. We do not have to use natural language for every application; some researchers claim that, in many cases, it is not the preferred mode of communication [Bates, Bobrow 84; Robertson, McCracken, Newell 81]. In natural language interfaces, the computer is the listener and the human the speaker. The listener's role is always more difficult, because he/she has to understand a problem based on the speaker's description. Our work has been primarily guided by the belief, that it is the user that is more intelligent and can be directed into a particular context. This implies that the essence of user-interface design is to provide users with appropriate cues. Windows, menus, spreadsheets, and so on provide a context (making the machine the speaker and the human the listener) that allows the user's intelligence to keep choosing the next step.

4.2 Expert Systems

We need a better understanding of the possibilities and limitations of expert systems research. We have to define the characteristics for problems which are suitable for expert systems research to generate realistic expectations. When we talk of a human expert, we mean someone whose depth of understanding serves not only to solve specific well-formulated problems, but also to put them in a larger context. The nature of expertise consists not only in solving a problem or explaining the results (which some expert systems can do to some extent), but of learning incrementally and restructuring one's knowledge, of breaking rules, of determining the relevance of something and of degrading gracefully if a problem is not within the core of the expertise.

These observations lead to the following research issues which have to be solved to make progress in expert systems research:

1. How to have an expert system learn from experience by changing its rules when it fails? The system may need to reorganize its initial knowledge over time and to reformulate the very basis of understanding of some topic as a result of new concepts and new experiences.

2. How to cope with unforeseen situations and adapt to failures; how to behave robustly in the face of error, the unexpected and the unknown? Experiments with human experts show that they are using mostly compiled knowledge (some rule-like mechanisms) if confronted with familiar examples but they exhibit a different mode of reasoning when confronted with unfamiliar examples; they go back to first order principles to supplement the rules. Expert systems research tries to capture this distinction by developing "deep models" compared to "shallow models" which are used to represent compiled knowledge.

3. How to incorporate models of the user and infer the user's goals and how to be able to understand the meaning of an event from a particular perspective?

4. How to exhibit self-awareness (to be used as the foundation of an explanation component and for the ability to recognize when a problem posed is outside its range of expertise)?

All these questions lead us back to research in Cognitive Science because there is no full understanding how human experts can cope with the problems described. We only can observe that they do.

Remarkably little has been done to evaluate expert systems. In the case of MYCIN [Buchanan, Shortliffe 84] one should be aware that it was never used routinely in patient-care settings. Despite the fact that it tried to be a "user-friendly" system (it could explain its reasoning, user aids were available (e.g. help, spelling correction and answer completion)), the developers claim that it failed along the following dimensions:

1. There was no recognized need by the individual practitioners for a system like MYCIN.
2. The system was not integrated naturally into the daily activities of practitioners (it was not what we called before an "embedded system").
3. A consultation may have taken 30 minutes to an hour.
4. The knowledge base became outdated through the existence of new medication.

The major shortcoming was in Human Engineering which led Shortliffe to the conclusion that *"Good advice is **not** good enough"*. The following interesting observations were made and the following dependencies were discovered (see chapter 32 in [Buchanan, Shortliffe 84]):

1. MYCIN did not use unconstrained English and it took new users some time to learn how to express themselves so they would be understood (an empirical justification of our claim made before).
2. To avoid natural language processing they used goal-directed (backward-chained) reasoning which allowed MYCIN to control the dialogue. This interaction style prohibited users from volunteering information which they may have considered relevant or from providing the system with a general description of a patient.

To overcome these shortcomings, the research team has worked on ONCOCIN (a program which assists physicians with the management of patients treated for cancer with chemotherapy). The system design of ONCOCIN attempts to prevent the computer system from being perceived as an unwanted intrusion into the clinic. To interact with the system, they have chosen a form-like approach which avoids the problem of natural language understanding (by supporting the role reversal between speaker and listener) and lets reasoning proceed in a data-directed fashion.

The DIPMETER advisor [Smith 84] is another expert system where the developers undertook an effort to critically review their achievements. Their findings have shown that the acceptance and real use of expert systems depends on far more than a knowledge base and an inference engine. They examined the relative amount of code devoted to different functions of DIPMETER and found that the *user interface portion was 42%* compared to 8% for the inference engine and 22% for the knowledge base.

To summarize these findings: expert systems should be considered as joint human-computer systems. Insights from Cognitive Science are required to overcome some of the limitations of current systems.

4.3 Cognitive Ergonomics

Research in cognitive ergonomics investigates and analyses the effect of technologies for human work and provides one of the foundations for user-centered system design. In the past properties of systems were investigated which could be measured with methods from physics (e.g. the design and layout of a keyboard). This approach is insufficient to evaluate systems in Artificial Intelligence and human-computer communication which try to support the human in decision making, planning, design and other cognitive activities. To evaluate these intelligent tools, criteria from Cognitive Science should play a crucial role. At the current stage of development, research in cognitive ergonomics should not be restricted to the comparison of finished products, but it should take an active part in the design and integration of cognitive dimensions in human-computer systems.

Contrary to hardware ergonomics, detailed prescriptions or check-lists cannot be provided for cognitive ergonomics, because so much is dependent on human cognitive abilities - how people behave, think and perceive the world. These abilities are not amenable being measured and predicted with the same precision that is possible with elements in the physical environment.

One of the major challenges in cognitive ergonomics research is to cope with conflicting design issues; some of them are:

- cognitive efficiency versus machine efficiency of a system,
- simplicity of a tool (to be easily handled by a large community) versus power (to be used in the construction of a large variety of different and complex systems),
- the necessity to remain compatible with existing systems (e.g. timesharing computers) versus exploiting the power of new media (e.g. networks of personal machines),
- ease for novice users (like mnemonic names in editors, menu-driven systems) versus convenience for experienced users (like a terse command language),
- tight integration between different subsystems versus flexible, reconfigurable modules or tool kits,

- systems which can be adapted by the end-users versus adaptive systems (which change their behavior based on profiles and models of the user).

5. Some Global Design Considerations of our Research

5.1 Symbiotic Systems

Symbiotic systems are based on a successful combination of human skills and computing power in carrying out a task which cannot be done either by the human or by the computer alone. We illustrate our conception of symbiotic systems by giving examples in different domains:

1. *Computerized axial tomography* (CAT scanning [McCracken 79]) is based on a cooperation between doctor and computer. The necessary inverse Fourier transformations involve an immense amount of computation and cannot be done without the help of a computer - and the interpretation of the data requires discrimination between subtle differences in density which is beyond current capabilities in image processing.

2. Kay [Kay 80] proposes a *symbiotic machine translation system* that is always under the tight control of translators. The system is there to help increase their productivity and not to supplant them. The fully automatic approach has failed badly in the past.

3. In *aircraft automation* [Chambers, Nagel 85] two different models are under investigation: the pilot's assistant and the electronic copilot which can be differentiated along the separation of tasks and control between humans and machine.

In our own work all intelligent user support systems (see section 5.2 and Figure 5-2) are best understood as symbiotic systems and are constructed this way. One rationale for symbiotic systems is to define and separate the strength and limitations of humans and computers (see Figure 5-1).

Symbiotic systems acknowledge the fact that most knowledge-based systems are intended to be of assistance to human endeavor and only a few are intended to be autonomous agents. A consequence of this is that a human-computer interaction subsystem is an absolute necessity. It should be understood that partially autonomous systems pose greater design challenges than fully autonomous systems would. The two agents have to keep each other informed about their decisions and actions and one of

The Human:

- has common sense,
- provides goals,
- defines subproblems and their relations to each other,
- uses previous experience (learns from mistakes),
- integrates knowledge from different sources,
- solves problems through analogy,
- selects suitable representation,
- copes with ambiguous, vague or uncertain environments,
- makes inductive decisions in novel situations,
- generalizes from analogous experience,
- improves and exercises judgement,
- converts sensory signals to cognitive symbols.

The Machine:

- has external memory support systems,
- provides levels of abstraction,
- maintains consistency,
- illustrates the consequence of our assumptions (with respect to previously defined rules),
- dissects complex information structures in different perspectives, (using filters),
- hides irrelevant information and does intelligent summarizing,
- supports visualization,
- allows UNDOs and supports an explorative and creative behavior,
- generates dynamic behavior according to static descriptions,
- takes care of low-level details,
- does not become complacent or yield to distraction,
- remembers large amounts of data without error,
- operates under real time constraints.

Figure 5-1: The Human Role versus the Computer Role in a Symbiotic Human-Computer System

the central questions is: who is in control in situations when there is a conflict in opinion? Knowledge-based systems develop their "own will" (or are encapsulations of the will of their designers and their understanding of the situation).

5.2 Intelligent User Support Systems

The "intelligence" of a complex computer system must contribute to its ease of use. Truly intelligent and knowledgeable human communicators, such as good teachers, use a substantial part of their knowledge to explain their expertise to others. In the same way, the "intelligence" of a computer should be applied to providing effective communication. In Figure 5-2 we illustrate a system architecture which we have developed in response to this design criteria. We have constructed a number of prototypical systems of the outer ring (e.g., our documentation system is described in [Fischer, Schneider 84], our help systems in [Fischer, Lemke, Schwab 85], and our critic is briefly described in section 7).

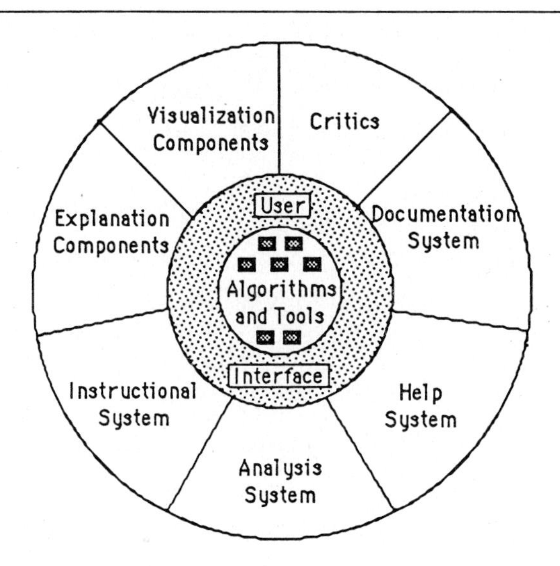

Figure 5-2: From Interactive to Intelligent Systems

To equip modern computer systems with more and more computational power and functionality will be of little use, unless we are able to assist the user in taking advantage of it. Empirical investigations [Fischer, Lemke, Schwab 85] have shown that on the average only a small fraction of the functionality of complex systems (e.g. UNIX, EMACS or LISP) is used.

It is insufficient for intelligent user support systems just to solve a problem or provide information. They need to do this in a way that the user can understand and question their advice. It is one of our working assumptions to hypothesize that learners and practitioners will not ask a computer program for advice if they have to treat the program as an unexaminable source of expertise. One has to provide windows into the knowledge base and into the reasoning processes of these systems at a level which is understandable by the user.

6. Examples of Tools

6.1 Tools for Visualization

Our goal is to build software components which allow us to take advantage of the power of the human visual system to provide insight and understanding instead of relying only on formal verification methods. We are in the process of integrating our visualization tools as components of an intelligent user support system.

The great potential of a computer system is that multiple representations can be generated *automatically* (i.e. without requiring the user to do much additional work) and *dynamically* (i.e. taking the actual work of someone into account). We will briefly describe KAESTLE [Boecker, Nieper 85] as a first step towards a *Software Oscilloscope* which is used to provide *explanations and illustrations* of complex processes and to generate explanatory materials within tutorial environments. The important point is that this can be done *on the fly*, i.e. explanations do not have to be precompiled and stored for later use, but instead can be generated dynamically when they are needed, using *actual* data.

With KAESTLE the graphical representation of a list structure (being the most important data structure of LISP) is generated automatically and can be edited directly with a pointing device. KAESTLE is integrated in our window system and has a menu-based interface (see Figure 6-1).

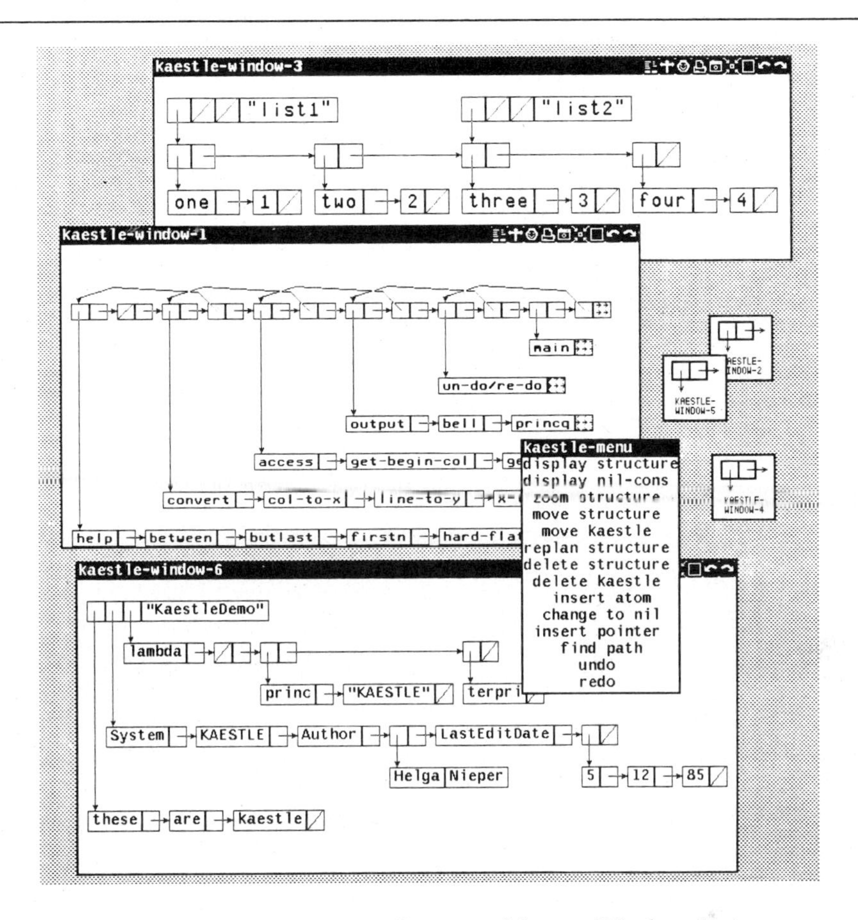

Figure 6-1: KAESTLE: Integrated in our Window System

KAESTLE is a valuable tool for the LISP beginner to understand certain aspects of the programming language which are difficult to explain otherwise (e.g. the difference between copying and destructive functions. Figure 6-2 illustrates the effects of **append** and **nconc**; the normal textual representation displayed in the "toplevel" window reveals no difference between the results of these two functions).

More experienced LISP programmers use it heavily to display and explore data structures which are difficult to represent symbolically, namely circular and reentrant structures (see "kaestle-window-1" in Figure 6-1).

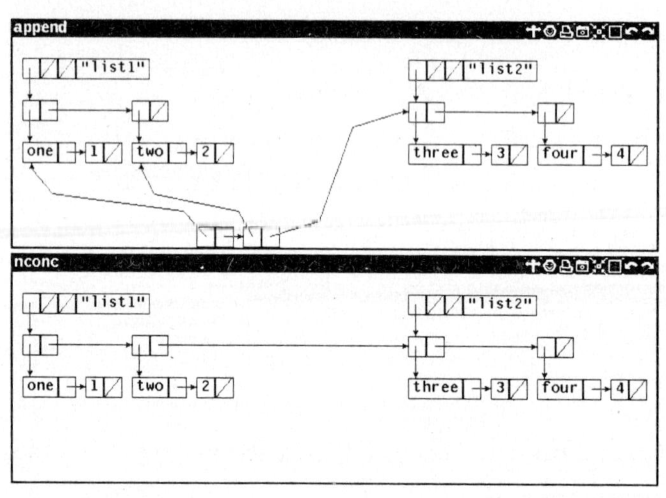

Figure 6-2: The Difference between Copying **append** and Destructive **nconc**

Providing a *software oscilloscope* will be even more important when computers become more intelligent; then we have to understand the complex internal processing (e.g. inference processes, inheritance networks) of knowledge-based systems. Without them we are left with "black boxes". The success of these systems depends on the designer's skill and artistic capabilities in choosing layouts, icons and graphical representations that are natural (i.e. convey the meaning of what they represent) and therefore easily understood.

6.2 Tools for Knowledge-Based Systems

The main issues in building knowledge-based systems are:

1. **Knowledge Acquisition:** How is knowledge acquired most efficiently from human experts and from data gathered by instruments? Can the experts themselves directly manipulate the knowledge base or do they need a knowledge engineer?

2. **Knowledge Representation:** How can the needed knowledge for complex problem solving processes be represented so it is effective for the inference engine and understandable for the human?

3. **Knowledge Utilization:** How can we retrieve the relevant knowledge needed in specific situations? Does the knowledge base help us in finding the relevant knowledge? Does it support browsing techniques to navigate through a knowledge space whose structure and content is unknown to the user in advance?

We have developed tools in all three areas based on a cognitive engineering approach. The main development has been ObjTalk [Rathke, Lemke 85; Rathke 86], an object-oriented, frame-based formalism for programming and knowledge representation. Based on ObjTalk, we have developed the graphically based knowledge editor ZOO [Riekert 86] and the browser LOOKAROUND [Rathke 86] (see Figure 6-3).

A major strength of object-oriented knowledge representations is their ability to provide a concise and intuitively appealing means of expression to a designer. This claim of intuitive appeal is based on our experience that object-oriented styles of description often closely match our understanding of the domain being modeled and therefore lessen the burden of reformulation in developing and understanding a formal description. Simon [Simon 81b] convincingly argues that complex systems take the form of a hierarchy and that hierarchical systems evolve faster based on the development of stable building blocks. Psychological research on memory has led to the conclusion that people faced with a new situation use large amounts of highly structured knowledge acquired from previous experience. This led to the notion of frames [Minsky 75] which have provided major design ideas for ObjTalk.

The implementation of our window, menu and icon systems (see Figure 3-1, [Boecker, Fabian, Lemke 85]) serves as a convincing example for these claims. The facilities of ObjTalk allowed for the highly modular and flexible structure of the window system and led to an open architecture providing high level interfaces to applications while at the same time keeping lower level parts of the system accessible for situations where nonstandard extensions are needed.

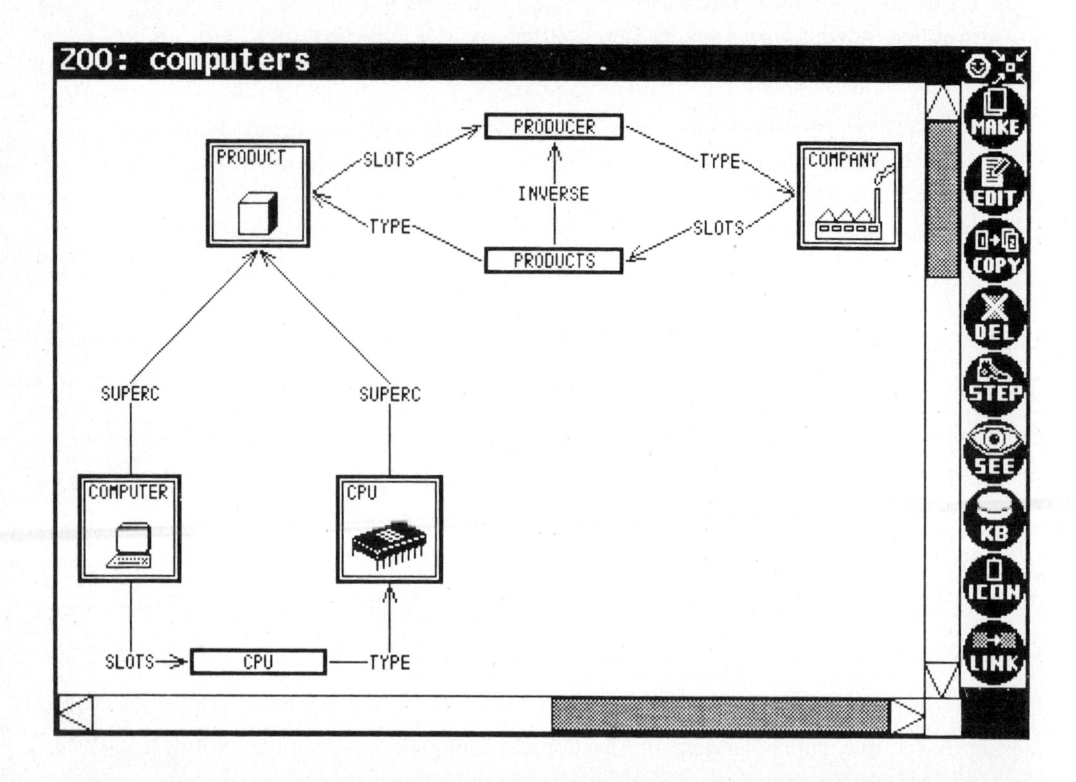

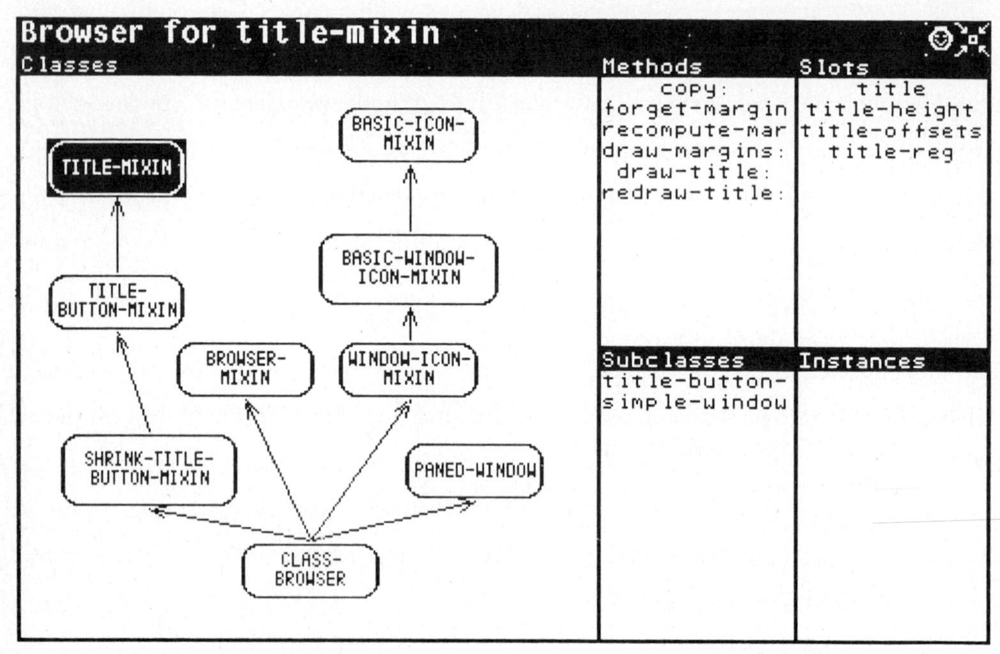

Figure 6-3: The Knowledge Editor ZOO and the Browser LOOKAROUND

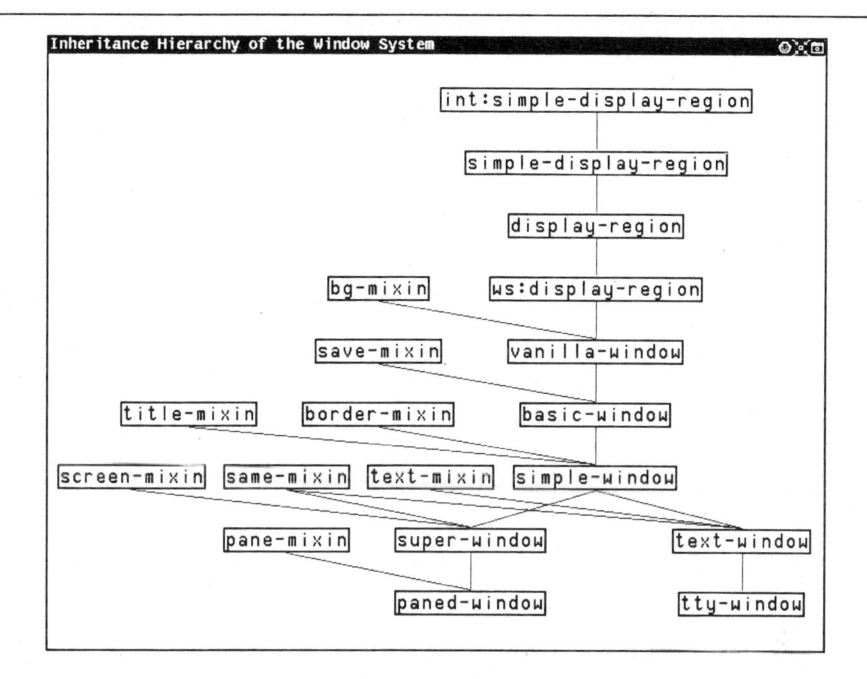

Figure 6-4: The ObjTalk Inheritance Hierarchy of our Window System

6.3 Interesting Questions from our Work for Cognitive Science

Cognitive Science in its current state is an empirical inquiry [Newell, Simon 76] in need of systems and environments which provide interesting phenomena. Our research over the last several years has generated a large number of interesting and challenging problems for Cognitive Science research. In the design of a truly two-dimensional interface, some of these question are:

1. With respect to windows:
   - How many windows should be on the screen at the same time?
   - Should they overlap or not?
   - Who determines their size and location (the system or the user)?
   - What kind of metatools (e.g. clusters which organize the spatial layout of icons; see Figure 3-1) are needed to make a complex window system manageable?
2. With respect to menus:
   - How do we associate menus with windows?

- What should be the maximum number of elements in a menu?
- What is the optimal ordering of elements in menus (e.g. alphabetical or functional)?

3. How do we catch the attention of the user (e.g. with blinking, color, multiple fonts, etc.)?
4. What is the primary interaction style; is it based on a *conversational metaphor* or a *direct manipulation style* [Norman, Draper 85]?

In our design and implementation efforts to construct intelligent user support systems (see section 5.2 and Figure 5-2), the following issues need a deeper understanding:

1. How can we support incremental learning processes and learning on demand [Fischer 86]?
2. What constitutes a good explanation (e.g., how can our *software oscilloscope* contribute to understanding and insight)?
3. Why is human advice so much more efficient than the advice which we can get from current user support systems?

The real difficulty to find answers to these questions is that we always have to consider a *multi-dimensional space* which includes the user (novice, casual or expert), the task (routine cognitive skill versus problem solving) and the technology (raster screen with mouse versus 24 by 80 character line terminal).

7. Critics: Examples of Cooperative Problem Solving Systems

One model frequently used in human-computer systems (e.g. MYCIN) is the *consultation model*. From an engineering point of view, it has the advantage of being clear and simple: the program controls the dialogue (much as a human consultant does) by asking for specific items of data about the problem at hand. The disadvantages are (as pointed out in section 4.2) that it prevents the user from volunteering relevant data and it sets up the program as an "expert", leaving the user in the undesirable position of asking a machine for help. We are in the process of developing a *critiquing model* which allows users to pursue their own goals and the program interrupts only if the behavior of the user is judged to be significantly inferior to what the program would have done.

The critiquing model will be used to support cooperative problem solving. When a novice and an expert communicate much more goes on than just the request for factual information. Novices may not be able to articulate their questions without the help of the expert, the advice given by the expert may not be understood and/or the advisee requests an explanation for it; each communication partner may hypothesize that the other partner misunderstood him/her or the experts may give advice which they were not explicitly asked for (the last aspect we have explored in our work on *active help systems* [Fischer, Lemke, Schwab 85]).

As a first instance of this class of systems we have constructed a critic for LISP called CODEIMPROVER [Fischer 86]. The system is used by two different user groups for two slightly different reasons:

- by intermediates who want to *learn* how to produce better LISP code. We have tested the usefulness of the tool by gathering empirical, statistical data using the students of an introductory LISP course as subjects.
- by experienced users who want their code to be "straightened out". Instead of doing that by hand (which these users in principle would be able to do) they use a system to carefully reconsider the code they have written. The system is used to detect optimizations and simplifications and it has proven especially useful with code that is under development and gets modified continuously.

The knowledge of the subject domain is represented in a network of interrelated concepts which represents the subject domain and the user has the ability to selectively browse through the knowledge easily. The system operates by using a large set of transformation rules that describe how to improve code. The user's code is matched against these rules and the transformations suggested by the rules are given to the user; the code is not modified automatically. The system explains and justifies its expertise while providing useful criticism and help to the user. It addresses several problems: users do not know about a function or tool, they do not know how to use it, they cannot modify the function or tool to their specific needs and they do not understand the results produced. The CODEIMPROVER addresses some of these problems as can be seen from Figure 7-1.

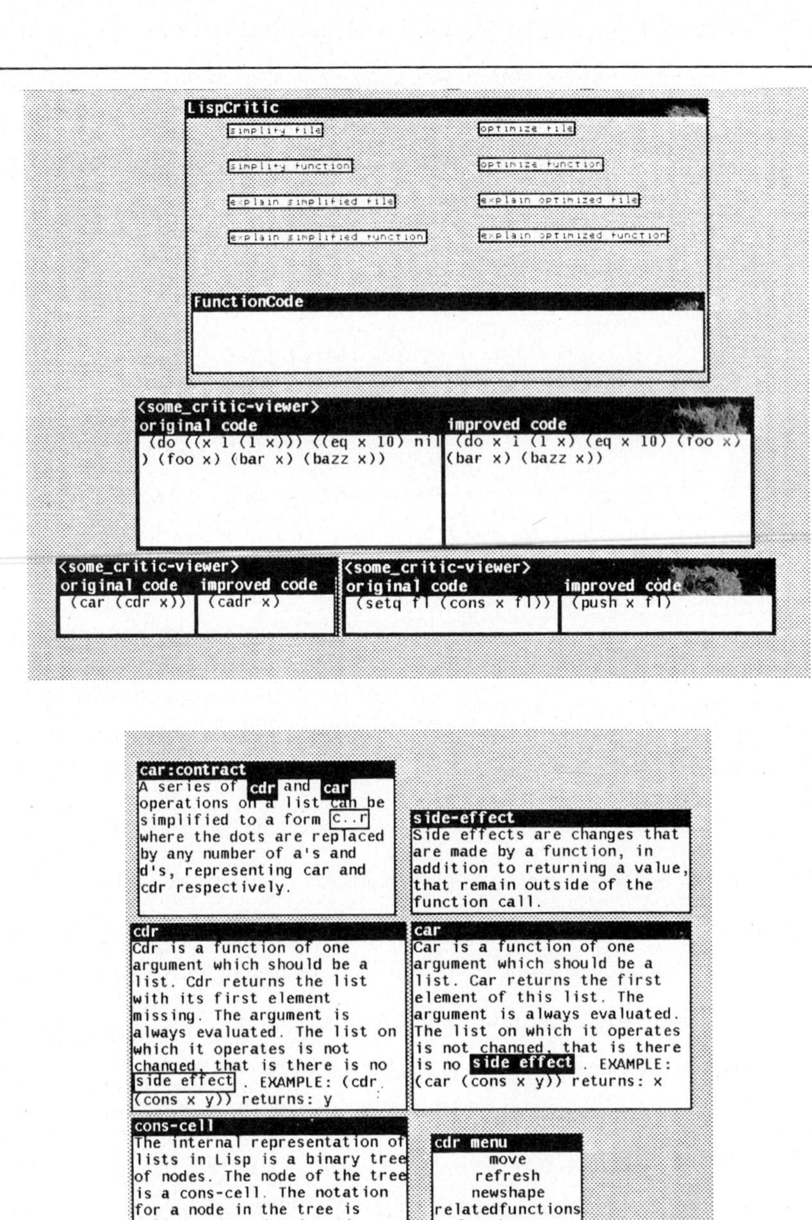

Figure 7-1: The CODEIMPROVER in Operation

8. Themes for the Future

Operational problems arise when complex systems are not designed with a full appreciation of the complex nature of human behavior and performance limits. We believe that in almost all technical systems, total automatic control will neither be technically feasible nor socially acceptable. This is why Cognitive Science and Cognitive Engineering is of crucial importance for the future success of Artificial Intelligence and other man-machine systems. A much broader understanding of design is necessary: if a human fails (in directing an aircraft, in using a computer system, etc.), then the persons who designed the systems should be included in the analysis of the failure. Under this perspective, "user error" is a controversial term: are the users casual agents (i.e. are they fundamentally at fault) or are they merely the last step in a sequence of events beginning with the design of the system.

To increase the effectiveness of human use of computers is a challenge which goes beyond technical questions. Computational power will be readily available. Today it may still be the case that one VAX is used by ten programmers, but the time that one programmer has ten VAXen is not far away. This raises the questions: what do we do with all this computational power? Who will be able to take advantage of it? Software must be designed to optimize not the way machines work, but the way people think. Computers will be (similar to television, telescope and microscope for our sight) amplifiers of human capabilities by expanding memory, augmenting reasoning and facilitating communication. One of the major challenges will be to resolve the design conflict between *shielding the complexity from the user* while *retaining a rich functionality.*

One way to achieve end-user control and end-user modifiability in highly complex systems is to advance *Human Computer Communication* to *Human Problem-Domain Communication* which permits us to build important abstract operations and objects of some given application area directly into the exploratory environment. This implies that the user can operate entirely in substantively meaningful abstractions. In most cases we do not want to eliminate the semantics of a problem domain by reducing the information to formulas in first order logic or to general graphs. Knowledge-based systems are the most promising way to achieve this goal. Human-Computer Communication should not be restricted to the construction of nice pictures on the screen. It is necessary to avoid that beautiful interfaces overshadow the limited functionality and extensibility of some systems. These observations provide the rational for one of the major research topics for the future: *how to bring knowledge-based systems and human-computer communication together to construct systems which are useful and usable.* The interplay between this research area and more general Cognitive Science research will be necessary and fruitful by providing design guidelines and challenging problems.

Acknowledgements

This paper has benefited from the critical remarks of my colleagues Catherine Cormack, Hal Eden, Andreas Lemke, Helga Nieper and Christian Rathke. They have also made major contributions to system components described in this paper. The research was supported by grants from the University of Colorado at Boulder and the Office of Naval Research.

References

[Bates, Bobrow 84]
M. Bates, R.J. Bobrow, *Natural Language Interfaces: What's Here, What's Coming, and Who Needs It*, in W. Reitman (ed.) *Artificial Intelligence Applications For Business*, Ablex Publishing Corporation, Norwood, NJ, 1984, pp. 179-194.

[Boecker, Fabian, Lemke 85]
H.-D. Boecker, F. Fabian, Jr., A.C. Lemke, *WLisp: A Window Based Programming Environment for FranzLisp*, Proceedings of the First Pan Pacific Computer Conference, The Australian Computer Society, Melbourne, Australia, September 1985.

[Boecker, Nieper 85]
H.-D. Boecker, H. Nieper, *Making the Invisible Visible: Tools for Exploratory Programming*, Proceedings of the First Pan Pacific Computer Conference, The Australian Computer Society, Melbourne, Australia, September 1985.

[Buchanan, Shortliffe 84]
B.G. Buchanan, E.H. Shortliffe, *Rule-Based Expert Systems: The MYCIN Experiments of the Stanford Heuristic Programming Project*, Addison-Wesley, Reading, MA, 1984.

[Chambers, Nagel 85]
A.B. Chambers, D.C. Nagel, *Pilots of the Future: Human or Computer?*, Communications of the ACM, Vol. 28, No. 11, November 1985.

[Fischer 84a]
G. Fischer, *Human-Computer Communication and Knowledge-Based Systems*, in H.J. Otway, M. Peltu (eds.) *The Managerial Challenge of New Office Technology*, INSIS Programme of the Commission of the European Communities, Butterworths, London, 1984, pp. 54-79, ch. 3.

[Fischer 84b]
G. Fischer, *Formen und Funktionen von Modellen in der Mensch-Computer Kommunikation*, in H. Schauer, M.J. Tauber (eds.) *Psychologie der Computerbenutzung*, Oldenbourg Verlag, Wien - Muenchen, Schriftenreihe der Oesterreichischen Computer Gesellschaft, Vol. 22, 1984, pp. 328-343.

[Fischer 86]
G. Fischer, *Enhancing Incremental Learning Processes with Knowledge-Based Systems*, in H. Mandl, A. Lesgold (eds.) *Learning Issues for Intelligent Tutoring Systems*, Springer Verlag, New York, 1986.

[Fischer, Lemke, Schwab 85]
G. Fischer, A. Lemke, T. Schwab, *Knowledge-Based Help Systems*, Human Factors in Computing Systems, CHI'85 Proceedings, New York, April 1985, pp. 161-167.

[Fischer, Schneider 84]
G. Fischer, M. Schneider, *Knowledge-Based Communication Processes in Software Engineering*, Proceedings of the 7th International Conference on Software Engineering, Orlando, Florida, March 1984, pp. 358-368.

[Kay 80]
M. Kay, *The Proper Place of Men and Machines in Language Translation*, Technical Report CSL-80-11, Xerox Palo Alto Research Center, October 1980.

[McCracken 79]
D. McCracken, *Man + Computer: A new Symbiosis*, CACM, Vol. 22, No. 11, 1979, pp. 587-588.

[Minsky 75]
M. Minsky, *A Framework for Representing Knowledge*, in P.H. Winston (ed.) *The Psychology of Computer Vision*, McGraw Hill, New York, 1975, pp. 211-277.

[Newell, Simon 76]
A. Newell, H.A. Simon, *Computer Science as an Empirical Inquiry: Symbols and Search*, CACM, Vol. 19, No. 3, 1976, pp. 113-136.

[Norman 81]
D.A. Norman, *Perspectives on Cognitive Science*, Ablex Publishing Corporation, Lawrence Erlbaum Associates, Norwood, NJ - Hillsdale, NJ, 1981.

[Norman, Draper 85]
D.A. Norman, S.W. Draper (eds.), *User Centered System Design, New Perspectives on Human-Computer Interaction*, Lawrence Erlbaum Associates, Hillsdale, NJ, 1985.

[Rathke 86]
C. Rathke, *ObjTalk: Repraesentation von Wissen in einer objektorientierten*

Sprache, PhD Dissertation, Universitaet Stuttgart, Fakultaet fuer Mathematik und Informatik, 1986, (forthcoming).

[Rathke, Lemke 85]
C. Rathke, A.C. Lemke, *ObjTalk Primer*, Technical Report CU-CS-290-85, University of Colorado, Boulder, February 1985.

[Riekert 86]
W.-F. Riekert, *Werkzeuge und Systeme zum Wissenserwerb und zur Wissensverarbeitung*, PhD Dissertation, Universitaet Stuttgart, Fakultaet fuer Mathematik und Informatik, 1986, (forthcoming).

[Robertson, McCracken, Newell 81]
G. Robertson, D. McCracken, A. Newell, *The ZOG Approach to Man-Machine Communication*, International Journal of Man-Machine Studies, Vol. 14, August 1981, pp. 461-488.

[Simon 81a]
H.A. Simon, *Cognitive Science: The Newest Science of the Artificial*, in D.A. Norman (ed.) *Perspectives on Cognitive Science*, Ablex Publishing Corporation, Lawrence Erlbaum Associates, Norwood, NJ - Hillsdale, NJ, 1981.

[Simon 81b]
H.A. Simon, *The Sciences of the Artificial*, MIT Press, Cambridge, MA, 1981.

[Smith 84]
R.G. Smith, *On the Development of Commercial Expert Systems*, AI Magazine, Vol. 5, No. 3, Fall 1984, pp. 61-73.

[Wilensky et al. 84]
R. Wilensky, Y. Arens, D. Chin, *Talking to UNIX in English: An Overview of UC*, Communications of the ACM, Vol. 27, No. 6, June 1984, pp. 574-593.

[Williams et al. 82]
M.D. Williams, F.N. Tou, R. Fikes, A. Henderson, T. Malone, *RABBIT: Cognitive Science in Interface Design*, Proceedings of the Cognitive Science Conference, Ann Arbor, Michigan, 1982.

[Winograd, Flores 86]
T. Winograd, F. FLores, *Understanding Computers and Cognition: A New Foundation for Design*, Ablex Publishing Corporation, Norwood, NJ, 1986.

KNOWLEDGE-BASED CONTROLLER FOR INTELLIGENT MOBILE ROBOTS

A. Meystel
Drexel University
Philadelphia, PA 19104, USA

Abstract.

An overview of the recent results in the area of Autonomous Control Systems is given in this paper. A concept of hierarchical nested controller is introduced, and an application of this concept is shown for control of intelligent mobile robots.

Key words: *Intelligent Control, Hierarchical Control, Cognitive Systems, Knowledge Based Controller.*

1. GENERAL CONCEPTS

Knowledge-based control of Intelligent Mobile robots is being considered as a part of a theory of **Autonomous Control Systems (ACS)** which is focused upon development of a theory, structures of algorithms, and design of systems for optimum motion of autonomous or semiautonomous systems operating in the **unstructured environment** with variable traversability of the state space (including not only the case of the obstacle strewn environment but any situation based upon incomplete and/or intrinsically imprecise information).

Autonomous Control Systems serve as a substitute for a human operator in the multiplicity of cases where **the danger** for a human operator is expected, and also in a number of cases where **human intelligent duties** require higher performance than is usually provided by a human operator. The structure of a typical Autonomous Control System in very general terms can be represented as shown in Figure 1. This structure consists of the closed loop (sensors, perception, knowledge base, control and actuation closed through the world), and an external connection via communication link which serves to assign a task, to receive the results of the reconnaissance, to abort the operation, to update the ACS, and also to provide communication of several ACS units working as a team.

Structures of hierarchical intelligent control [G.SARIDIS, 1977, 1983] are potentially proper tools for solving this type of problems if they are given at least a rudimentary capability of performing **cognitive operations** which is usually done by techniques of artificial intelligence, self-organizing automata, and neural networks. This paper concentrates upon a subset of cognitive operations associated with **motion planning for autonomous intelligent mobile robots.**

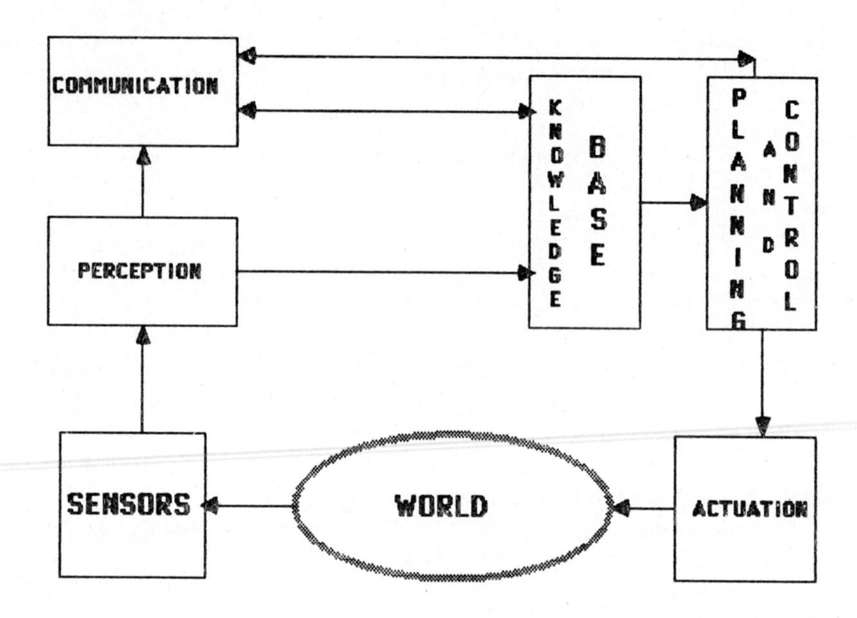

Figure 1. Structure of Autonomous Control System

Planning is traditionally considered as a process which is being performed separately from the process of control. This is acceptable for the vast multiplicity of systems where planning can be performed off-line, and the process of control can be started given a number of **unchangeable goals.** This decoupling upper levels of decision-making (or off-line stages) from the lower levels of decision-making and immediate performance (or on-line stages) is probably the most characteristical property for distinguishing planning and control stages of operation as well as subsystems of any device where **constant human involvement** is presumed.

The problem of motion planning was given substantial attention in the literature on AI and robotics. However, problems of **optimum planning** as well as **optimum control** until now do not have applicable solutions. Motion planning is frequently understood in the context of **"solvability" of the problems of positioning or moving** the object rather than in the context of finding the appropriate location and/or determining the desirable trajectory of motion. Nevertheless one cannot argue that the real problem of concern is finding the location and/or trajectory of motion which provides a desired value of some "goodness" measure (e.g. the value of some cost-function).

The emphasis of the well known concept of the "configuration space approach", is done upon techniques of constructing the **admissible** swept volume but no **optimality** is considered, and certainly, no dynamics of the motion is discussed. Most of the algorithms based upon the theories mentioned above are oriented toward **off-line operation**, they require considerable time and constant human involvement. Finally, all of the existing works presume complete knowledge of the environment, and operate in a structured world. No result is known contemplating planning of motion in unstructured situation. In the meantime, this situation is a typical one for a hierarchical system of **intelligent autonomous control** which are expected to be in the focus of **our research.**

Reduction of the motion planning to a pure geometric issue can be understood given complexity of this problem and the mathematical elegance of solutions it generates. We would like to express here our appreciation of the results containing the advancements in using configuration space (T. LOZANO-PEREZ, M.A. WESLEY, W. RED, H.V. TRUONG-CAO. A.A. PETROV, T.M. SIROTA), in finding the minimum distance path under geometrical constraints (J.Y.S. LUH, C.S. LIN, L.A. LOEFF, A.H. SONI, S.M. UDUPA, C.E. CAMPBELL),upon the network (G. GIRALT, R. SOBEK, R. CHATILA, V.A. MALYSHEV), using the Voronoi diagrams for motion planning with and with no retraction (R.A. BROOKS, C.K. YAP) as well as introduction and the treatment of such problems as "moving the ladder", "moving the piano", and so on (C.K. YAP). Various methods of minimum path construction have been applied based upon determining the "potential field" surrounding the obstacles (O. KHATIB), global flow analysis using Gauss-Jordan elimination (R.E. TARJAN), applicable when the full knowledge of the world is presumed to be given.

The following comments should be taken in account: the above mentioned works reflect a paradigm of **off-line static planning of motion trajectory in a cluttered limited space.** Clearly, this is only a part of the whole problem- an important one but just a part. As soon as the **on-line real-time** planning is required, as soon as **dynamics** is involved, as soon as the "plant" is **complex and hierarchical** one, and the world is **not uniform** and not well known, finally, as soon as the computer power turns out to be limited (as happens in all autonomous systems) - then the old premises are not working anymore. The **whole problem** must be solved based upon new set of premises pertaining to **autonomous systems** and using other means of solution. The new premises generate the new promises, and the new strategies of planning are being devised.

Planning is understood as a process of determining the desirable motion goals and/or trajectory without actually moving. Thus, planning is expected to generate the input to the control system in the form of a description of the state,(or the sequence of states) to be achieved during the operation. This means, that the system of planning must actually **predict** the motion trajectory which should be admissible, and at the same time it should provide the desirable value of the cost-function. This also means that the input for control system is to be determined as a result of planning. Finally, it means that **planning and control should be considered as a joint process and/or system because of their intrinsic interactive character and mutual influence.**

On the other hand, this **prediction** should be obtained before the actual motion started, and information on the world at this stage is usually incomplete. Thus, the **contingencies** must be contemplated based upon construction of **plausible situations** for which the uncertain variables and parameters should be estimated. Clearly, the role of control subsystem is presumed to be a **compensatory** one, so that the uncertainties of the initial information, and the inconsistencies of the cost-function formulation could not diminish the expectations about the desirable results of motion conveyed to the control system in the form of the plausible situations.

The motion trajectory which is obtained as a result of planning must be considered as a task of control, and is to be given as an input to the controller. Thus, the better is the result of planning (i.e. the better the uncertainties have been handled at the stage of planning, and the closer the preplanned trajectory is to the optimum control trajectory), the easier will be the **compensatory role** of the conventional **controller** which is presumed to be at the bottom of the planning-control hierarchy.

The following postulates represent the problems of planning and control, at least for robotic application.

Postulate of Planning. At the stage of planning the optimum trajectory of the motion (for a system) is to be determined. Optimum trajectory is meant to minimize, maximize, or provide the definite value for the cost-function, and this condition must be satisfied no matter how the actual control operation is being provided, and by which means.

Postulate of Control. At the stage of control the preplanned trajectory must be performed (by a system) in such a way that the deviation of the motion from the preplanned trajectory and/or the deviation of the cost function from the preplanned value, would be minimized or would not exceed the desired value. If the desired value was exceeded the plan must be recomputed with new information taken in consideration (**replanning**).

So, the operations of **planning and control are complementary**, and the better the planning is done, the easier is the control operation.

Since the available accuracy of the world description at the stage of planning is limited, the problem appears of determining the limit of resolution accepted from the system of **world representation**, or the **knowledge base.** In order to limit the required volume of computations, the **nested world description** is considered to be appropriate as a general concept for constructing the knowledge base. The idea of nested world description is illustrated in Figure 2.

It is presumed that each level of the hierarchy has a limited computing power, and it is our intention to utilize the total power for decision making. (Certainly, there is another problem here: whether each of the levels does require the **maximum computing power** for decision making. However, the problem of proper distribution of computing power among the planning-control levels, is not being addressed in this paper). Then, at the upper level, where the resolution is the lowest, the broadest view of the world is taken. Each subregion of the "planner's view" can be **zoomed down** to the "navigator's" level of consideration.

One can see that for a smaller fraction of the world a multiplicity of details is emerging immediately. A reduced partial problem has totally different, and very rich contents. same thing happen when the zooming down to the "pilot's" level is being considered. The scope is being reduced, although the number of details has increased. One can see also that at the lower level, even the direction of the IMR motion is different which does not contradict to the main goal of the overall operation.

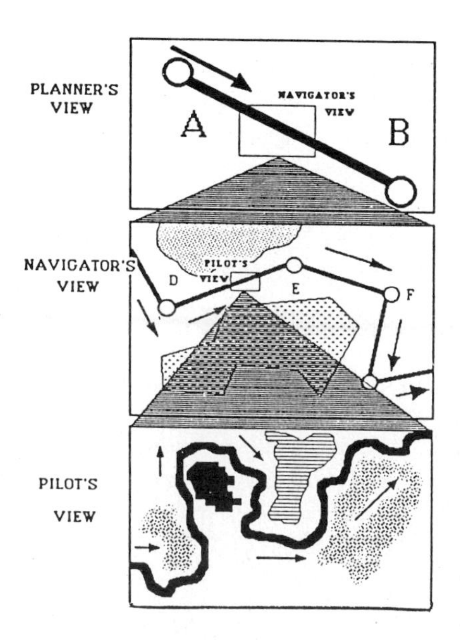

Figure 2. Nested hierarchy of the world views

Conventional control theory employs very rich and productive world description based upon analytical and computational methods rooted in the mathematics of differential and integral calculi. These methods should be substantially enhanced and modified when the hierarchical systems of intelligent control are under consideration (G. SARIDIS, Y. HO, Y. PAO, J. ALBUS, all of the sources at 1985). In fact, a new methodology must be developed based upon dealing with linguistic representation of the word (G. SARIDIS,1984), and the well known analytical and computational constructions based upon differential and integral calculi are expected to be just a particular case of the new methodology utilizing a **knowledge based world representation** (J. MYLOPOULOS, H.J. LEVESQUE, P.J. HAYES, R.J. BRACHMAN, all of the sources at 1984).

Knowledge based world representation (KBWR) matches perfectly all of the approaches existing in the theory of intelligent control since the hierarchical organization of information is an intrinsic property of KBWR. Knowledge base employing the principle of nested world description, allows for the performance of planning within a desirable part of the **context** at a definite level of resolution with a definite **accuracy**, and with a satisfactory **volume of computation**.

Then, the reduced **area of interest** can be assigned as a fraction of the previously considered part of the context, and the same limited computing power can be applied to the smaller part of the initial world description dealing with information at a higher level of resolution and providing the results with higher accuracy. KBWR systems can store the rules obtained from experience, can decompose the analytical world representation into systems of rules, and can generate new rules based upon processes of **clauses generation in the version spaces.**

Thus, it is implicitly presumed that the system ("plant") is being controlled by a **nested hierarchical controller** (known also as a "crew" Planner-Navigator-Pilot) introduced in according to the methodology of design of intelligent controllers (A. MEYSTEL, 1984). This controller is shown in Figure 3. The structure of the nested controller corresponds to the structures of ACS in Figure 1 and of the world view in Figure 2. Knowledge Base is named "Cartographer" here since in IMR the major part of Knowledge Base can be thought of as a system of terrain maps. Planning-control system is shown as a hierarchical production system. Interlevel feedbacks are represented as a hierarchy of reporters.

It does not employ the conventional analytical description of the dynamic system: it is substituted by an equivalent linguistical representation with fuzzy evaluations on quantitative data (L.ZADEH, 1983,1984). Clearly, the nature of the autonomous system to be controlled is of no importance here: the results are presumed to be equally applicable to the domain of multilink manipulators as well as to the area of autonomous mobile robots, autonomous machines working in a hazardous environment, or other systems and/or technologies using autonomous processes.

From the above conceptual sketch of the **unified AI based, planning/control theory** one can see that within the autonomous system, the knowledge based AI procedures for planning do permeate substantially through the structure of the intelligent control system, especially when the goal of optimum operation is posed. It is clear from the above description that an organic merger between the control part and the AI part including the system of representation as well as a number of systems which constitute the so called "**inference engine**" and can be considered as some enhancement of the standard AI concept of **production system.**

This production system should be based upon a **concept generating knowledge base (CGKB)** which is capable of self-clustering procedures upon the raw information from the sensors, and then organizing this information (knowledge) in task-oriented hierarchies. Then, using this information within a **planning-control** dual environment of the intelligent controller would be natural since the sequence of generalizations was done upon the perceptual data arriving from the environment. This is one of the few ways available that promise some practical capability of decoupling the autonomous machine from the human being given the complexity of environment and the problems to be solved by a machine. This detached (from communication with humans) system of control must have all of the above mentioned properties unified under the name of **artificial cognition.** Controllers with these properties, or **cognitive controllers** have a capability of joint autonomous planning/control (JAPC) operation.

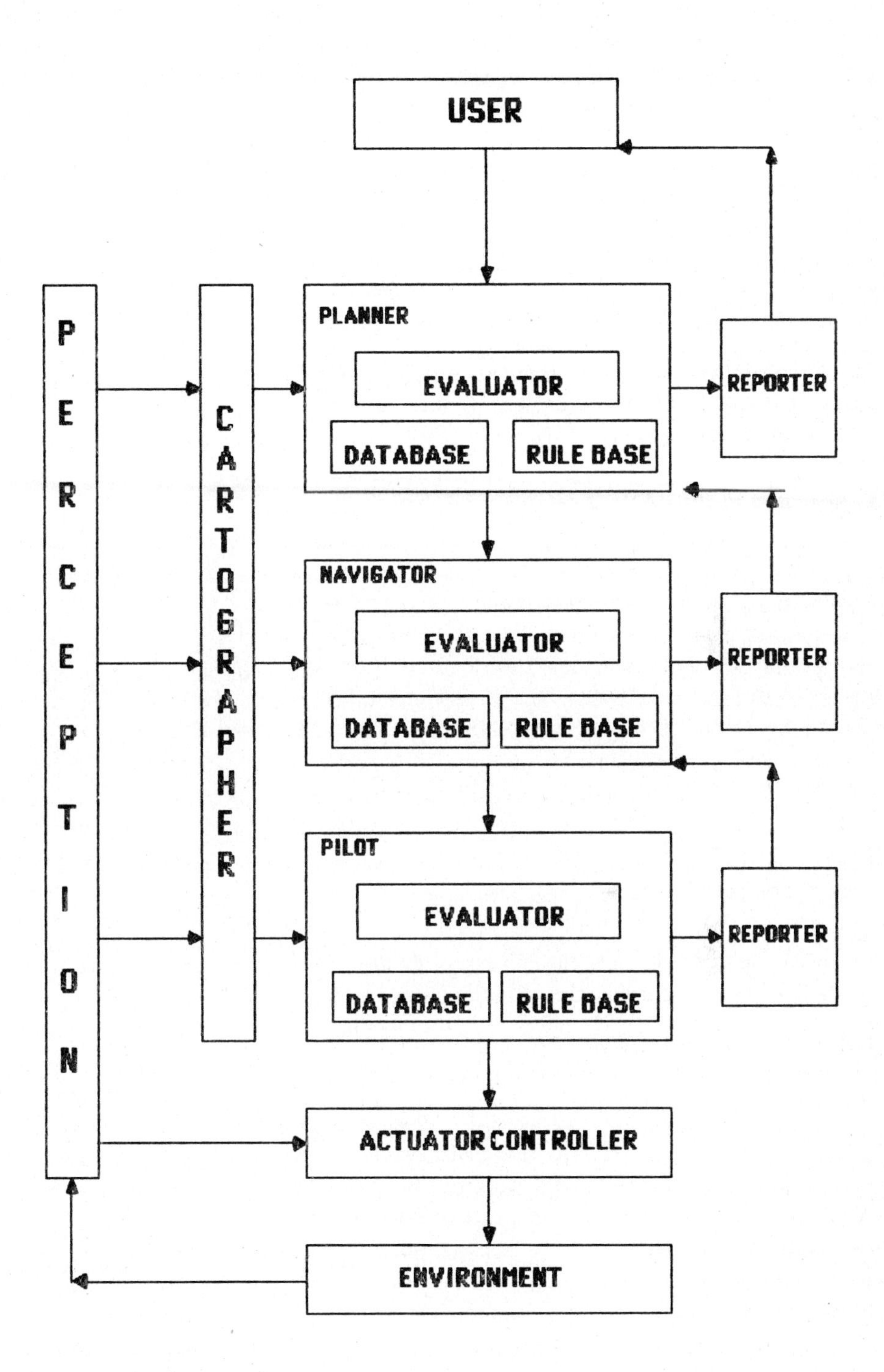

Figure 3: General Structure

The following directions of research are of the further interest:

I. Development of a Unified Optimum Planning/Control Theory based upon Knowledge Base World Representation.

II. Development of a Theory of Optimum Autonomous Control Oriented Reasoning upon Knowledge Bases.

III. Development of a Theory of Joint Autonomous Planning/Control (JAPC) upon Knowledge Structures.

IV. Theoretical and Experimental Analysis of the Self-Organizing Knowledge Bases fed by the Information from Autonomous Robotic/Manufacturing Perceptual Systems.

V. Learning in the JAPC Systems upon Knowledge Structures.

2. Hierarchical Nested Controller

This section illustrates how the theory of ACS is being implemented in a real device: Intelligent Mobile Robot (IMR). The structure of Intelligent Controller is being decomposed into the three levels of control: **PLANNER, NAVIGATOR,** and **PILOT.**

2.1 Subsystems of the Hierarchical Nested Controller

Based upon the mission formulated, modes of operation, and the technology presumed, the controller for IMR has the following subsystems:

1. The mechanical assembly including the platform, drive train, the subsystem of energy conversion and transmission, the actuator subsystem for acceleration, coasting and braking, the subsystem of steering, the subsystem of mounting, and providing mechanical motion to the sensors. (In the future, the subsystems of manipulator arms will be added).

2. The system of motion control which is defined as a combination of principles, algorithms and devices destined to change the configuration of path, speed trajectory, the law of accelerating and decelerating, etc. in a way which will minimize the cost-function of the operation. Thus, the motion control operations include:

- **planning** the overall path which contains the rough description of the motion trajectory, speed trajectory, etc. as functions of time for a large period of time, with low resolution of the description and low accuracy of the result;
- **navigation** over the results of planning which is intended to increase the resolution of the mentioned trajectories, reduce their error by using the additional information which is not

available for the planner;

- **piloting,** i.e. implementing the results of navigation; piloting transforms the prescribed trajectories into more specified and detailed sequence of maneuvers, including those not available for the algorithms of navigation since the resolution of the path and speed description, as well as the accuracy, is very high (additional spontaneous data are utilized.)

- **execution control** (conventional control), which is based upon the need to follow the results of piloting as accurately as this is achievable at the existing level of informedness of the parameters of the elements of the system of energy conversion and transmission.

3. System of sensors ("perception" of IMR autonomous controller), which includes all technological means of observing the real motion and judging on the closeness of the motion to the results of the prescribed and desirable control, and also the technological means of proper transformation and/or integration of the measurement results.

4. System of knowledge which is required:

4.1 For the mechanical assembly - in order to utilize properly the ideas implanted in this assembly by the design procedure.

4.2 For the motion control system - in order to determine the IMR assignment, to synthesize a solution concerned with the motion trajectories at all of the considered levels of control: planning, navigation, piloting, and compensation of deviations (conventional lower level controller).

4.3 For the system of sensors in order to interpret the results of sensing (e.g. scene understanding), and to properly assign the results of observations to the different levels of the hierarchy of motion control.

The interrelation between these functional structures can be illustrated by Figure 4,a. The hierarchical character of these subsystems is illustrated in Figure 4,b which makes these interrelations more understandable. It is clear that the same bulk of knowledge at each level of the hierarchy serves simultaneously to the needs of determining the trajectory of motion (control), and to the needs of image understanding (scene interpretation). This is illustrated in Figure 4,c.

Thus the "sensor" hierarchy performs the monitoring of motion", the "knowledge" hierarchy performs the "organizing the results of monitoring", and the "motion control" hierarchy provides the "control of motion".

"Planning" is a control of motion at the level of strategic decision making: the overall goal of the operation is synthesized on the basis of initial sensor information (the optimum trajectory of motion is determined), and the rough trajectory of motion is sketched.

"Navigation" controls the motion at the tactical level. After the motion has started the new information is perceived, the trajectory of motion is made more precise.

"Piloting" controls the motion at the operative level. This includes any vision based decision-making processes performed in "real time" on-line.

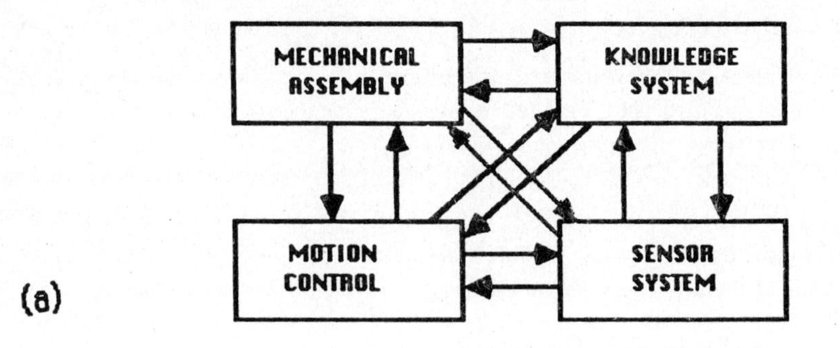

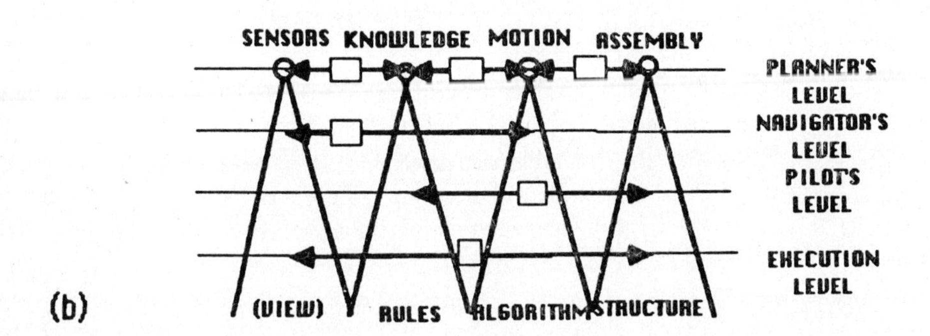

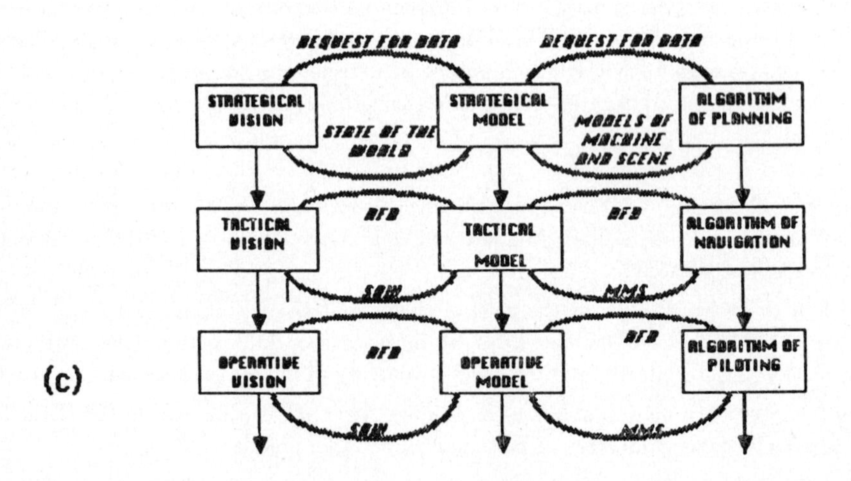

Figure 4. Functional subsystems of IMR (a) and their nested hierarchies (b) reflected as interrelationships between the hierarchies of "vision", "world models", and "planning-control" (c).

Finally, the **"Execution Control"** transforms the sequence of commands from the Pilot into the actual motion. One can see that the "Execution Control" is also a "real time control" however it includes the part of "rendering" not a decision making anymore.

The higher is the level of hierarchy the more is the lag in time between the motion control determined and actually executed. At the upper level we have full "non-real time control" of the motion. The degree of lagging is illustrated in Figure 5 as a possible way of temporal reasoning.

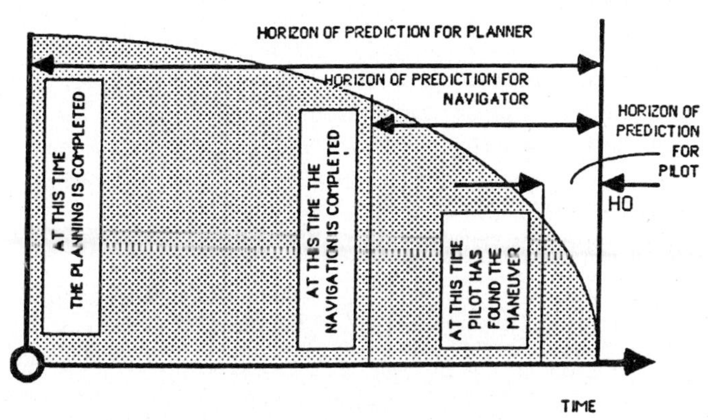

Figure 5. Temporal reasoning in the nested hierarchy.

2.2 Operation of the Subsystems

2.2.1 Motion Control

a) Planning Operations . Given the mission objectives and description, the IMR control system will have to plan a set of motion operations to achieve the objectives. This will require the analysis of the goal location, analysis of the terrain and thematic map data, analysis of constraints, preferences, expert information, and all mechanisms of descriptive and mathematical models to identify a plan of operation. World information will be interpreted to define milestones and "passage points" through which the mechanisms must operate. These milestones and the list of passage points correspond to the strategic areas that should be traversed to achieve the goal. The strategic areas may include any descriptive entity of interest such as "passageway entrance","crater","rock","edge of the auxiliary mechanism", etc. If most of the space is passable (though, in different degree) the milestones of the future path should be determined at the level of "planning".

Thus, the process of planning is determined as finding the passage points and milestones which should be achieved sequentially during the IMR operation.

Since in most of the real situations there is a multiplicity of path alternatives, the following cost-function should be taken into consideration:

$$\text{COST-FUNCTION} = \text{TOTAL TIME OF MOTION} + \text{RELIABILITY VIABILITY} + \text{ENERGY LOSSES}$$

The actual computational formula cannot be written for the whole cost-function. However, some rough estimations can be given for each of the components of the cost function. Indeed, the components of the total time of motion should be based on the estimation of time required to execute motion from one milestone to another. Thus, the time will depend on distances and the average speed that is achievable within the region to be traversed. Energy consumed and the values of reliability and viability should be estimated in the same way.

Traversability is a complex function of slope, ground cover, tactical considerations, including the mechanisms specifications, and other factors. Traversability affects the time of operation, energy (fuel) consumed and both the reliability and viability. The information of traversability should be obtained from the system of knowledge representation which combines the information of the conditions of repair including the description of the craters, properties of terrain given in the initial map, and the communication between the components of the functional structure is determined.)

Since the points of passage and the other milestones are given in a fuzzy manner, and since the cost function is presented also imprecisely (sometimes in a form of heuristical prescriptions), the mathematical model for the "planning" in most of the cases cannot be defined and the knowledge based reasoning system should substitute the accurate computations. In a case of limited energy, one of the factors to be taken in consideration is so called "resource allocation", e.g. the proper distribution of energy among the parts of the motion trajectory. After "the best chain of the milestones" is obtained (and a couple of competitive alternatives is stored), the routes between the passage points should be generated. The final planning function is to identify operations to be performed when the goal position has been achieved.

b) Navigation Operation The function of the navigation process in the generation of a route also (with possible alternatives) for the system or its particular part to traverse from a milestone to a milestone. Associated with these routes will be the behavior alternatives. Such behavior alternatives will include the control mode and the maneuver specifications (e.g. travel at a definite rate of speed and a definite rate of fuel consumption), sensor selection strategies (e.g. don't turn on an active sensor unless absolutely necessary), and exception event responses (e.g. how to behave when the enemy is encountered). One can understand that the alternatives of behavior while moving from a milestone to a milestone are determined as a result of reasoning using the knowledge of the navigation level.

Navigator obtains the world representation from the knowledge system

(at the navigator level of the hierarchy of knowledge) and uses this representation to generate routes between the passage points. This can be a decomposition process, i.e. a recursive generation of sub-passage points of sub-milestones until at some level the traversability of the space is analyzed. However in practice, navigator cannot determine more than just one sub-milestones: the updating information is required. Thus after the first intermediate point of the possible route between the milestones is found, the motion should begin, and the next sub-milestone will be determined after the information is updated.

c) Piloting. PILOT receives from the NAVIGATOR the location and the direction of the first milestone, the desired rate of speed and the required rate of the energy consumption, and it must generate a sequence of commands.

Piloting is a conditional process that depends on the current state of the environment. For example, the commanded speed of the mechanism must depend on actual surface conditions, sudden turn, etc. Pilot control decisions will have to be implemented that can accommodate this conditional process. Maneuers of the road should be reflected in correctives introduction, stones and puddles should be taken in consideration, typical maneuvers of traversing the narrow passageway, entering the gate, drastic direction changes at a small patch of ground - all of these sub-problems should be determined from the information of higher resolution and using the knowledge and the reasoning power of the pilot level.

d) Execution Control. Finally the Execution Control subsystem is even more near-sighted than PILOT. It tries to perform the system of PILOT commands precisely and its sensors observe the speed of actuators, torques in the shafts, acceleration in the system, tilts, oscillations and so on. For the Execution Control subsystem of the system of Motion Control, this requires more than just transmit the upper level commands to the lower level performer. The "World" should be taken in account now at the highest level of resolution and attention to the details.

Control monitoring includes: plan monitoring (PM), route monitoring (RM), maneuver monitoring (MM), and execution monitoring (EM), and is being performed by the system of "reporters" (which can be considered as a generalization of observers for knowledge-based controllers).

PM) During execution of a plan, the system will be required to continually reevaluate its plans with regard to the currently inferred situation in order to determine if the process is succeeding or failing. In the event that the process is failing, plan inadequacies must be identified and some degree of replanning should be initiated. Then the milestones should be determined again. In some cases the rejected alternatives of milestones can be utilized.

RM) During the trajectory execution, a new information can determine the need in repeating the procedure of navigation. In some cases the rejected alternatives of routes can be utilized.

MM) During the low level controller operation the inability is revealed to follow the sequence of commands generated by PILOT. Another maneuver must be assigned. Clearly, the process of monitoring is fully perception (sensor system) dependent.

EM) Is operating as a conventional controller feedback.

2.2.2. Sensor System (Perception)

Perception System includes subsystem of Sensors and subsystem of processing the information, intended to monitor the motion control from the strategies, tactical and operative views. The perception subsystem provides the ability to extract information from the sensor data in order to construct an internal model of the world. The perception subsystem will use apriori models of the environment provided by the knowledge base, and will output its world model into the knowledge base. Thus the sensor subsystem (perception) contains two major components: sensors "per se", and the system for processing their output. The processing presumes dealing with knowledge stored in the system of knowledge and reasoning upon this information until the judgment concerning with the scene is obtained (see Figure 6).

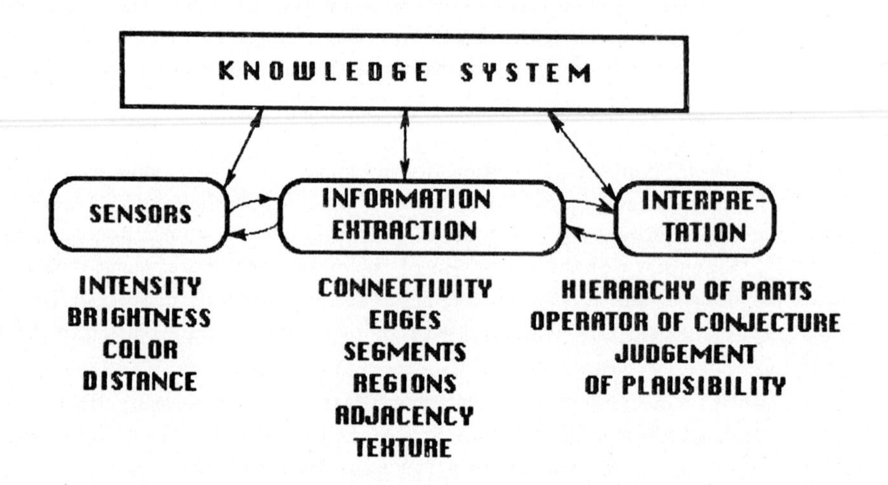

Figure 6. Links between perceptual system, and knowledge base.

a) Sensors. The sensors provide quantitative measures of observables in the environment. The selection of sensors is tailored to the mission requirements and the IMR system should, therefore, have the flexibility to accept any advanced sensor component relevant to the cluster of operations within any combat engineering mission.

b) Information Extraction. A judgment of the scene can be made when the direct sensor output is transformed in some elements which help to make conjectures and try to verify them in the context. Thus the connectivity edges, segments, regions should be found and roughly organized in a matrix of adjacency.

c) Information Interpretation. This component of the Perception subsystems makes a model of the environment to construct observed scene using the tool of reasoning, making conjectures. The information processing architecture should be separated into high level and 'reflecting' processes similar to human perception.

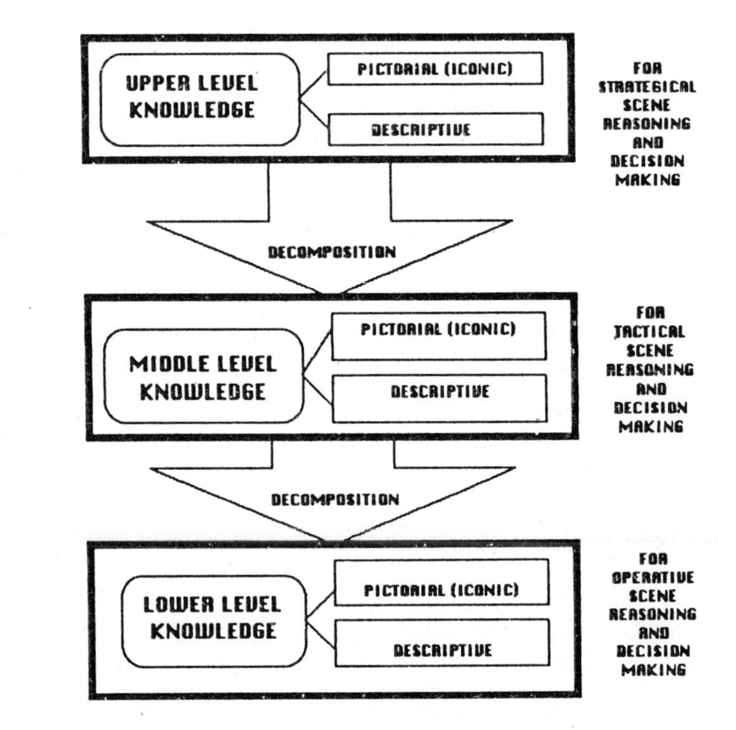

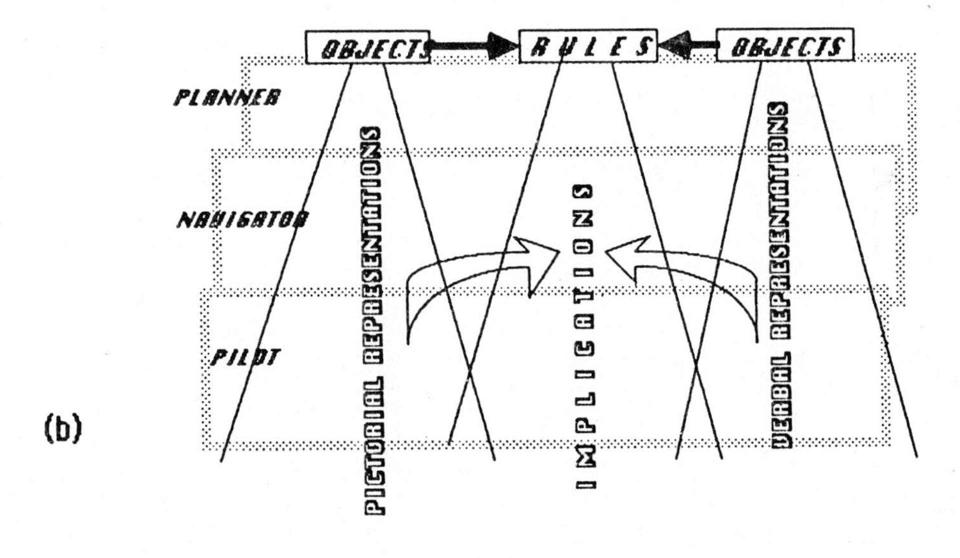

Figure 7. Structure of dual world representation (a) and the hierarchy of clauses connecting the pictorial and descriptive hierarchies (b).

Low level models of critical components in the environment such as surfaces and obstacles are essential for the pilot and execution control to maintain control of IMR. Furthermore, this information is needed rapidly or the system will never achieve the desirable effectiveness. More detailed models of the environment representing complex relationships of features in the environment are needed for higher level decision making. Image understanding approaches that model a large number of features, their relationships, and attributes will not achieve the rates required for real time control at velocities specified by the demonstrations required in the near future.

2.2.3. Knowledge System

From Figures 4,c and 6 one can see that the hierarchies of Motion Control System and Sensor System (Perception) are both linked to the body of knowledge and via this body they are linked together. Thus, the knowledge structure can be visualized as shown in Figure 7,a for a case with dominating visual perception. At each level of the Knowledge Hierarchy, the knowledge of the level is separated into the following three areas:

a) World Iconic (Pictorial) Information which contains two groups of information: visualized iconic images, and transformed iconic images ,e.g. making different projections or substituting them by symbols (photopictures and topographical maps belong to this group).

b) World Descriptive Information which in turn is divided into two groups: presented in natural or limited natural (commander) language, and resented in symbolic language code.

c) Rules (implications) which include the collection of "IF-THEN "clauses organized in hierarchies according to the hierarchy of problems determined by the IMR mission: the hierarchy of IF-THEN clauses contains all rational knowledge of the world and of the mechanisms including the whole vehicle and the subsystems on- ,and off-board. The above mentioned clauses are being obtained from the linguistical and pictorial networks using a number of formal procedures.

Thus having two hierarchies of world knowledge (Figure 7,b), the third hierarchy contains all "IF-THEN" knowledge concerning all objects and their decompositions which could be visualized in the world (or verbally described). Subsequently, the hierarchy of rules contains also the models of physical laws, which are to be utilized in reasoning, model of the vehicle, its subsystems, and the environmental models.

The IMR knowledge bases presume the property of "self-awareness". Indeed, the world description (e.g. topographical map) contains also the information of the present location in the memory of its previous path plans and expectations concerning the future activities of IMR. In a sense the world information can be divided into 3 parts: a world scene representation, an image modeling knowledge, and the motion control knowledge. World scene representation model s the IMR controller's view of its external environment (Figure 8).

A single integrated scene representation is required, that incorporates both apriori terrain information and sensor derived information collected during the IMR operation. A terrestrial scene model will be utilized that also contains a representation of the uncertainty or ambiguity about the system's knowledge of the scene.

The world scene representation has two important subsets: the model of environment, and the knowledge of the surfaces. In turn the model of environment can be separated into several areas. First, the environment must be observed through the use of active or passive sensors. The selection of sensors and coverages must ensure that the features being modeled are completely observed in the data. Second, information must be extracted from the raw data. This extraction process will be a dynamic function of the mission and the apriori knowledge of the environment.

The information must be sufficient for the decision making functions. For example, if the system is only interested in reaching one point of the road from another, the "expectation", or the roughly averaged edge of this road may be the adequate information. However, if the system works in a complex tactical environment it might be necessary to extract detailed information of the environment such as the nature, and the relations of features to one another. The final area of environmental modeling is the creation of the model from the extracted information.

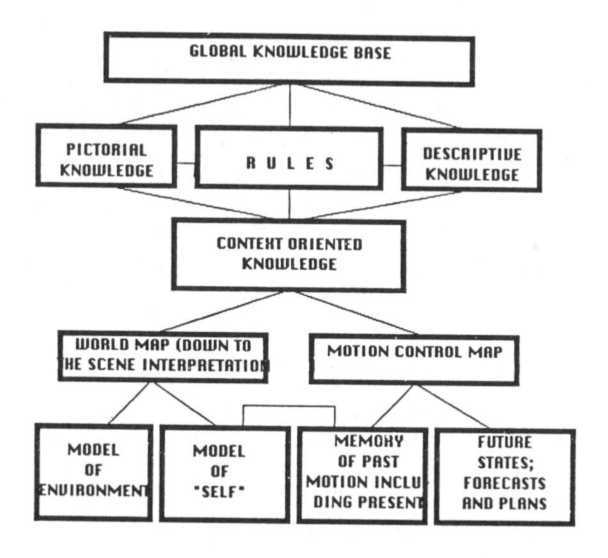

Figure 8. Functional structure of the IMR knowledge base.

This model will relate the observed features to a prior knowledge of similar features. For example, if a small bush is detected, a prior knowledge may indicate that the vehicle may travel over the bush without considering it to be an obstacle. The model will also relate features to each other in terms of their location, and logical relationships. The knowledge of

Terrain (Terrain Knowledge Base-TKB) is a primary resource for path planning and landmark recognition.

The TKB contains both grid-based data (digital topographic maps, for example), and symbolic data. TKB contains rules and operations for transforming terrain data into forms needed by the other high-level vision modules. For example, in a landmark identification task, TKB would output successively more detailed symbolic descriptions of landmarks until a unique one was identified.

Image Modeling knowledge contains models of all visualized elements required by the vision system for effective obstacle detection and analysis, landmark finding and terrain feature segmentation and typing. Typical features for which general models will be developed include rocks, bushes, roads, edges of the road.

Motion Control knowledge provides for the information of the previous, present, and future motion (plans and forecasts). The Motion Control system will continuously generate, update or modify, the controls at each level for vehicle movement. Alternative (contingency) plans will be developed and ranked as to their appropriations. This knowledge base will likely be in a hierarchical tree form in which branches represent plan alternatives, or the depth of the tree represent levels of plan refinement down to the routes maneuvers and speed and acceleration trajectories. The structure of Knowledge System which reflects the above mentioned structuring is shown in Figure 8.

The Global Knowledge Base (GKB) is being stored as a hypergraph of correspondence between two networks: pictorial and descriptive labels. The links between the subgraphs carved out of the hypergraph according to the task produce clauses (rules). GKB is being decomposed into Context oriented Knowledge Base which can be decomposed into the World Map (down to the scene interpretation) and into the Motion Control Map. The first one is a "snapshots" representation (noun-world). The second is relation among snapshots belonging to the different moments of time - temporal knowledge.

3. Pilot's Level of IMR Knowledge-based Controller

Pilot works under two inputs: windshield view (a hierarchy of statements which represent the present state of the world, $\mathbf{S_p}$), and the Navigator's command (a hierarchy of statements which represent the desired state of the world, $\mathbf{S_d}$). Windshield view should be understood in real time, and the command (output hierarchy of statements which represent commands **A** for providing the desired actuation, or motion to be performed, $\mathbf{\Delta M_a}$) should be generated for the lower level controller. Thus, the following holds

$$\mathbf{S_d - S_p = \Delta M;\ S_p x \Delta M = A,}$$

which represent the set of operations computing the hierarchy of motions to be performed, and look-up table for determining which command should be applied. Together, these

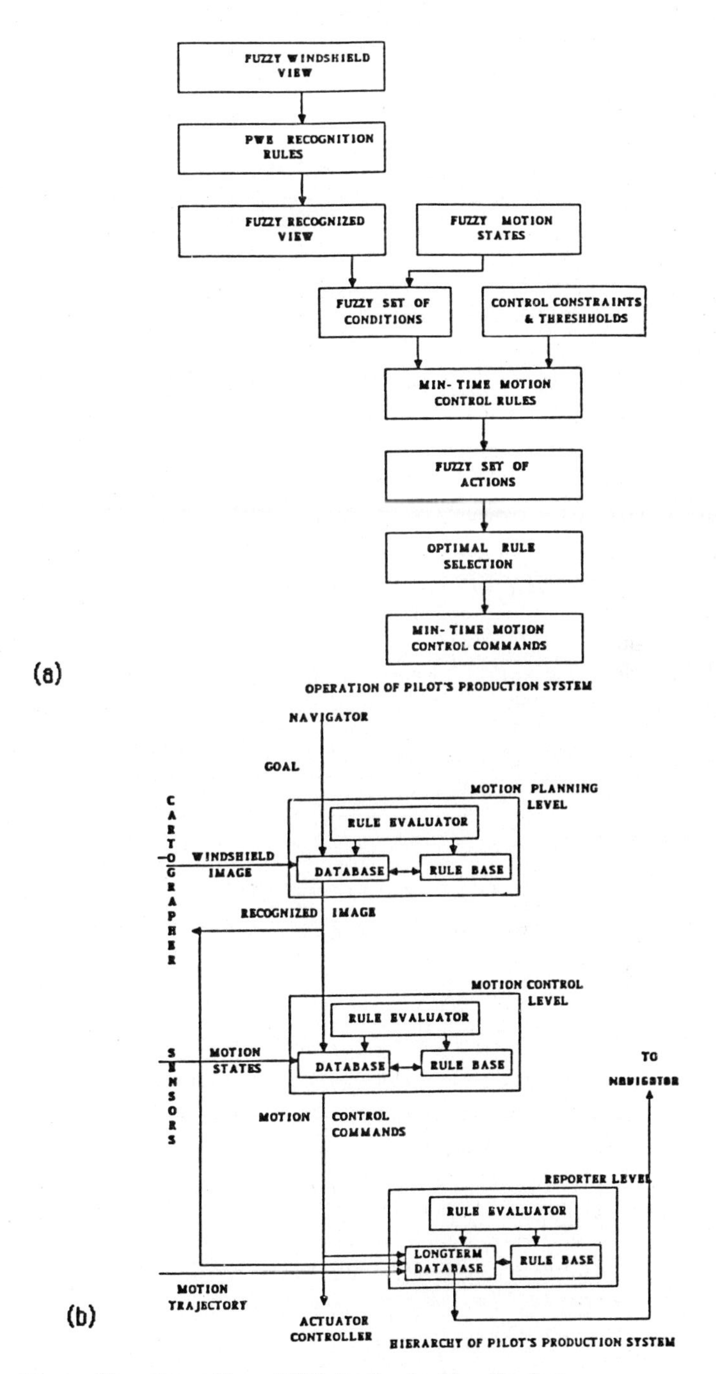

Figure 10,a. Operation of Pilots Production System
Figure 10,b. Hierarchy of Pilot's Production System

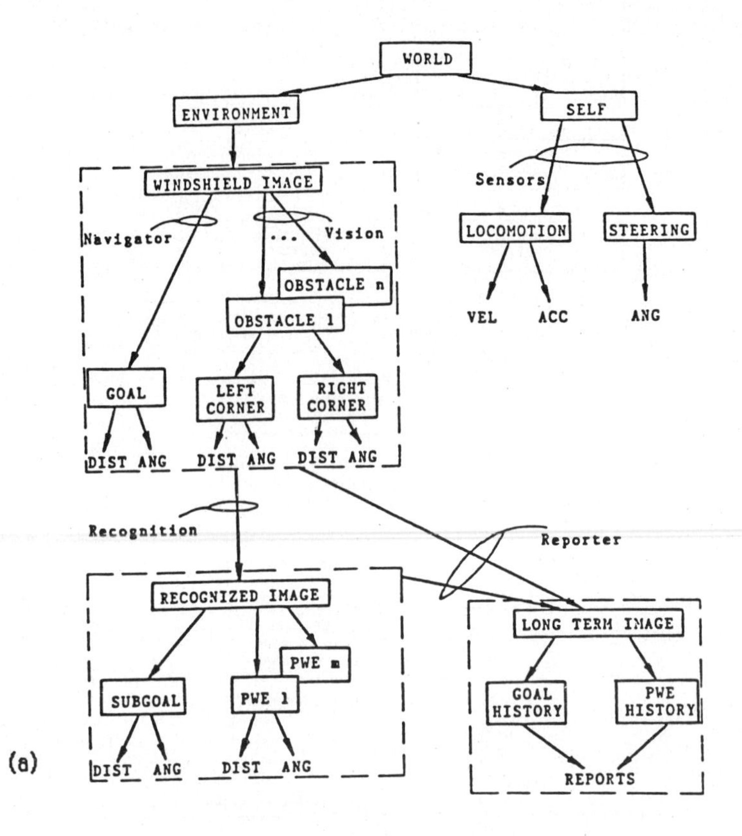

(a)

DISTANCE (DIST)

DIST ε { VERY CLOSE (VC), CLOSE (C), MEDIUM CLOSE (MC),
MEDIUM FAR (MF), FAR (F), VERY FAR (VF),
TOO FAR (TF) }

ANGLE (ANG)

ANG ε { ±VERY NARROW (VN), ±NARROW (N), ±MEDIUM NARROW (MN),
±MEDIUM WIDE (MW), ±WIDE (W), ±VERY WIDE (VW),
±OUT OF SCAN (OOS) }

VELOCITY (VEL)

VEL ε { VERY SLOW (VS), SLOW (S), MEDIUM SLOW (MS),
MEDIUM FAST (MF), FAST (F), VERY FAST (VF),
TOO FAST (TF) }

(b)

Figure 11. Hierarchy of Pilot's Vocabulary (a), and list of linguistic variables (b)

operations, and the look-up table can be consideres as the pilot's **inference engine.**

Operation of this inference engine is illustrated in Figure 10, a and b.. The structure of Pilot's vocabulary is shown in Figure 11,a, and a list of examples of linguistic variables is given in Figure 11, b.

This Pilot is implemented in Drexel Dune-Buggy, (an intelligent mobile robot created in the Laboratory of Applied Machine Intelligence and Robotics at Drexel University). The results of motion simulation for a variety of maps is given in Appendix. Additional information can be obtained from the references: C. ISIK, 1984; E. KOCH,1985; A. MEYSTEL,1985; A. MEYSTEL,1986.

References.

J.S. Albus, C.R. McLean, A.J. Barbera, M.L. Fitzgerald, "Hierarchical Control for Robots, and Teleoperators", Proc. of the Workshop on Intelligent Control, Troy, NY, 1985

R.J. Brachman, "On the Epistemological Status of Semantic Networks",in *Associative Networks: Representation and Use of Knowledge by Computer*, ed. by N.V. Findler, Academic Press, New York, 1979

R.A. Brooks, "Solving the Find-path Problem by Good Representation of Free Space", Proc. of the 2nd AAAI Conference, Pittsburgh, 1982.

R. Chavez, A. Meystel, "Structure of Intelligence For An Autonomous Vehicle", Proc. of the IEEE Int'l Conf. on Robotics and Automation, Atlanta, GA, 1984.

G. Giralt, R. Chatila, M. Vaisset, "An Integrated Navigation and Motion Control System For Autonomous Multisensory Mobile Robots", 1st Int'l Symp. of Robotic Research, 1983

G. Giralt, R. Sobek, R. Chatila, "A Multilevel Planning and Navigation System For a Mobile Robot: A First Approach to Hilare", Proc. of IJCAI-79, Vol. 1, Tokyo, 1979

P.J. Hayes, "The Logic of Frames", in *Frame Conceptions and Text*

Understanding, ed. by D. Metzing, Walter de Gruiter & Co, Berlin, 1979

Y.C. Ho, "System Theory, and Operation Research - A New Fusion of Mathematical Modeling and Experimentation ", Proc. of the Workshop on Intelligent Control, Troy, NY ,1985

C. Isik, A. Meystel, "Knowledge-based Pilot for an Intelligent Mobile Autonomous System", Proc. of the First Conference on Artificial Intelligence Applications", Denver, CO, 1984

C. Isik, A. Meystel, "Structure of a Fuzzy Production System for Autonomous Robot Control", Proc. of SPIE, vol. 135, *Applications of Artificial Intelligence III*, ed. by J. Gilmore, Orlando, Fl, 1986

M. Julliere, L. Marce, H. Place, "A Guidance System For a Mobiel Robot", Proc. of the 13th Int'l Symposium of Industrail Robots and Robots 7, Vol. 2, "Future Directions", Chicago, IL, 1983

O. Khatib, "Real-Time Obstacle Avoidance For Manipulators and Mobile Robots", (Preprint, 1985)

E. Koch, C. Yeh, G. Hillel, A. Meystel, C. Isik, "Simulatio of Path Planning For a System With Vision and Map Updating", Proc. of IEEE Int'l Conf. on Robotics and Automation, St. Louis, MO, 1985.

L.A. Loeff, A.H. Soni, "An Algorithm for Computer Guidance of a Manipulator between Obstacles", ASME Transactions, paper 74-DET-89, 1971

T. Lozano-Perez, "Spatial Planning: A Configuration Space Approach", IEEE Transactions on Computers, Vol. C32, No. 2, 1983.

T. Lozano-Perez, M.A. Wesley, "An Algorithm For Planning Collision-Free Paths Among Polyhedral Ostacles", Communications of ACM, Vol. 22, No. 10, 1979.

J.Y.S. Luh, "A Scheme For Collission Avoidance with Minimum Distance Traveling For Industrial Robots", Journal of Robotic Systems, 1(1), 1981.

J.Y.S. Luh, C.E. Campbell, "Collision-Free Path Planning for Industrial Robots", Proc. of the 21st IEEE Conference on Decision and Control, vol. 1, Orlando, FL, 1982.

J.Y.S. Luh, C.S. Lin, "Optimum Path Planning For Mechanical Manipulators", Trans. of the ASME, J. of Dynamic Systems, Measurement and Control, Vol. 102, June, 1981.

V.A. Malyshev, "Representation of An External Medium and Planning Construction of Program Motions of A Manipulator", Engineering Cybernetics, No. 3, 1981.

L. Marce, M. Julliere, H. Place, "An Autonomous Computer-Controlled Vehicle", Proc. of the lst Int'l Conf. on Automated Guided Vehicles, 1984.

A. Meystel, "Intelligent Control of a Multiactuator System", in *IFAC Information Control Problems in Manufacturing Technology 1982*, ed. by D.E. Hardt, Pergamon Press, Oxford, 1983

A.Meystel, A. Guez, G. Hillel, "Minimum Time Path Planning for a Robot", Proc. of the IEEE Conf. on Robotics and Automation, San Francisco, CA, 1986

A. Meystel,*Primer on Autonomous Mobility,* Drexel University, Philadelphia, PA, 1986

J. Mylopoulos, H.J. Levesque, "An Overview of Knowledge Representation", In *On Conceptual Modelling,* ed. by M.L. Brodie, et al, Springer-Verlag, New York,1984

Y.-H. Pao,"Some Views on Analytic and Artificial Intelligence Approaches", Proc. of the Workshop on Intelligent Control, Troy, NY, 1985

A.A. Petrov, T.M. Sirota, "Obstacle Avodance by a Robot Manipulator Under Limited Information About Environment", Automation and Remote Control, No. 4, 1983

W. Red, H.V. Truong-Cao, "The Configuration Space Approach to Robot Path Planning", Proc. of the 1984 ACC, Vol. 1, San-Diego, CA, 1984

G.N. Saridis, *Self-Organizing Control of Stochastic Systems* , Marcel-Dekker, New York, 1977

G.N. Saridis, "Intelligient Robotic Control", IEEE Transactions on Automatic Control, Vol. AC-28, No. 5, 1983

G.N. Saridis, J.H. Graham, "Linguistic Decision Schemata for Intelligent Robots", Automatica, Vol. 20, NO 1, 1984

G.N. Saridis, "Foundations of the Theory of Intelligent Control", Proc. of the Workshop on Intelligent Control, Troy, NY, 1985

R.E. Tarjan, "Fast Algorithm for Solving Path Problems", Journal of the ACM, Vol. 28, No. 3, 1981

S. Udupa, "Collision Detection and Avoidance in Computer Controlled Manipulators", Proc. 5-th IJCAI, Cambridge, MA 1977

C.K. Yap,"Algorithmic Motion Planning", in *Advances in Robotics,*v.1, ed. by J.T. Schvartz, C.K. Yap, Lawrence Erlbaum Publ.,1985

L.A. Zadeh,"Fuzzy Sets", Information and Control, Vol.8, 1965

L.A. Zadeh,"PRUF-a Meaning Representation Language for Natural Languages", in *Fuzzy Reasoning and its Applications,* ed. by E.H.Mamdani, B.R.Gaines, Academic Press, 1981

L.A. Zadeh, "Fuzzy Probabilities", Information Processing and Management, vol.20, NO.3,1984

Appendix

RESULTS OF MOTION SIMULATION

Example 1 (Figure A-1,a) The objects are initially unknown, the triangle in the lower left hand corner of the figure symbolizes the IMR initial position and orientation. The cross in the upper right corner, is the goal. The path followed is shown by the sequence of little circles. The density of circles is related to the velocity trajectory (the denser the circles are, the slower is the vehicle). In this example, PILOT works with no Navigator involvement. There are no small scale obstacles.

The initial result of Pilot motion leads IMR in a "bay" within the second large obstacle. At the point "a" Pilot changes the IMR direction since the wall ahead is already being perceived. New change in direction is commanded at point "b" when the upper wall has been visualized. IMR behavior within the circle "c" shows that it does not change its direction immediately, and that the fuzzy linguistic controller is not precise enough.

Finally, at "d", a new change of direction is performed because the third obstacle has appeared in the IMR field of view. At all linear parts of the path the speed trajectory is as should be expected; IMR accelerates, then moves at a maximal speed, then brakes as soon as the turn should be done. This can be seen more grafically in Figure A-1,b where instead of circles, the triangles are used to depict the motion.

Example 2. (Figure A-1,c) In this example, the Navigator is added. It computes the shortest path which turn out to be ABCD. Interestingly enough, the beginning of motion is shaped in the same peculiar way as it was in the pictures Figure A-1,a and b.Analysis has indicated that because of the nature of the Navigator's algorithm it shows the corner of the obstacle as the first subgoal. For the Pilot's rules it is not a good choice since the width of IMR is reflected in the rules and the vehicle can collide with the wall. So, it commands to turn right in order to keep the limit of safety, and then returns to the navigated course.

Area 'b' demonstrates the program failure: Pilot failed to remember the next subgoal, and did not bother to ask Navigator for the renavigation. So it decided to travel to the goal with no supervisor guidance, and of course soon it saw that the obstacle is ahead. The 'wall following property' returns IMR on its preplanned trajectory. Since the vertex C is approached now from another angle, the path at the interval CD is done with overrun, in a different way than it was done in the case Figure A-1, a and b.

Example 3 (Figure A-1,d). Little squares are the obstacles which are visible for the Pilot but not for Navigator. Thus these obstacles avoidance is performed only when they appear in immediate closeness from IMR.

Example 4 (Figure A-2,a).Operation of the system is explored under one condition changed: the maximum level of speed is increased with no change in the maximum values of acceleration and deceleration. Clearly, the system is working near the limits of operational safety.

Example 5. (Figure A-2,b) The small objects-squares are initially unknown. Note that the three adjacent small obstacles makes the PILOT change the IMR path, since the PWE (passageway entrance) that existed in the previous examples is now completely blocked. One can see that at the interval AB the motion is waggling (Figure A-2,c) not just slow as is shown in Figure A-2,b. This can be explained by the fact that the motion is being done almost with no gains in performing the task of becoming closer to the **GOAL**.

Example 6. (Figure A-3,a) Object is unknown, initial position of the IMR is in the upper left corner, goal is in the lower part of the picture, PILOT and NAVIGATOR are working together. Pilot is confused: it follows the Navigator's direction but the obstacle preclude IMR from reaching the goal. The "wall following" leads IMR farther and farther from the goal. Than, it starts a new attempt to follow Navigator's directions, and again with no result. This is a **trap**.

Example 7. (Figure A-3,b) Interesting example of program failure. After completion of two circles in the same direction the motion of IMR become erratic: the program was not capable of dealing with angles more than 2π.

Example 8 (Figure A-3,c) Here the operation is shown after the program was fixed, and the problem of this type of trap was solved as follows. If there is a **loop repetition** in the history of IMR motion which is recorded by Reporter, then a new obstacle is being placed in the map which makes this particular PWE nonexistent anymore.

Example 9 (Figure A-3,d). The upper PWE is totally closed. In this case, the wall following strategy creates a loop then **reporter** eliminates this PWE and the total path is being renavigated at the Navigator's level.

Example 10 (Figure A-4,a). This is the experience of IMR simulation in a new map. In this case, IMR is first intended to achieve the goal moving in the direction of arrow 1. However, the PWE AB turned out to be locked. Wall following strategy brought IMR beyond the map borders (due to the programming error the map happened not to be locked everywhere. So, IMR managed to achieve the goal using the forbidden space outside the map. Ruthlessly, the surpassing passage was also blocked (Figure A-4,b). Since it happened before the **loop repetition** rule was introduced, the system again did what was shown in Figure A-3,b and used the advantages of the **erratic behavior**.

Example 11 (Figure A-4,c) Otherwise coping with the circumstances was explainable and consistent. Dotted lines are the Navigator's plans. Little squares are the small obstacles which are visible only at the Pilot's level. Then Pilot avoids these obstacles properly. Similarly it operates when the passageway AB turns out to be totally closed (Figure A-4,d).

Example 12 (Figure A-5,a-d). This simulation illustrates the importance of adjusting the numerical values in the rule clauses for determining the different character of IMR motion. In Figure "a" Pilot does not allow any risc in doubtful cases. If the passageway **seems** to be not sufficient in its width, another direction is commanded. Sure enough, it leads to substantial loss in productivity when the situation repeats frequently. By assigning the different values to the discrimination statement of the program, the "**pessimistic pilot**" from "a" is being gradually transformed into the "**optimistic pilot**".

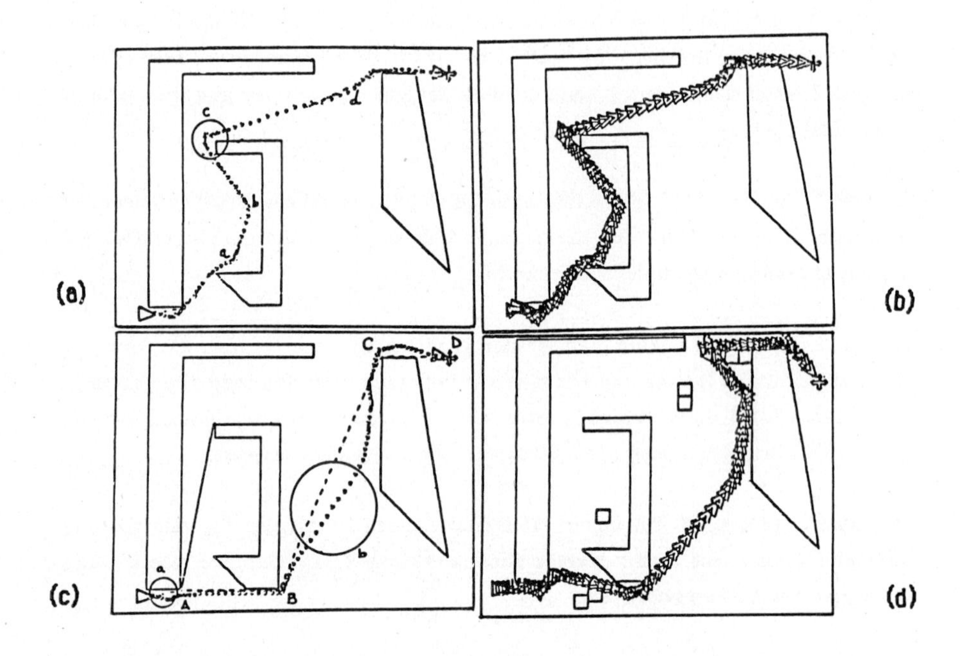

Figure A-1. IMAS operation in Polygonal World (set 1)

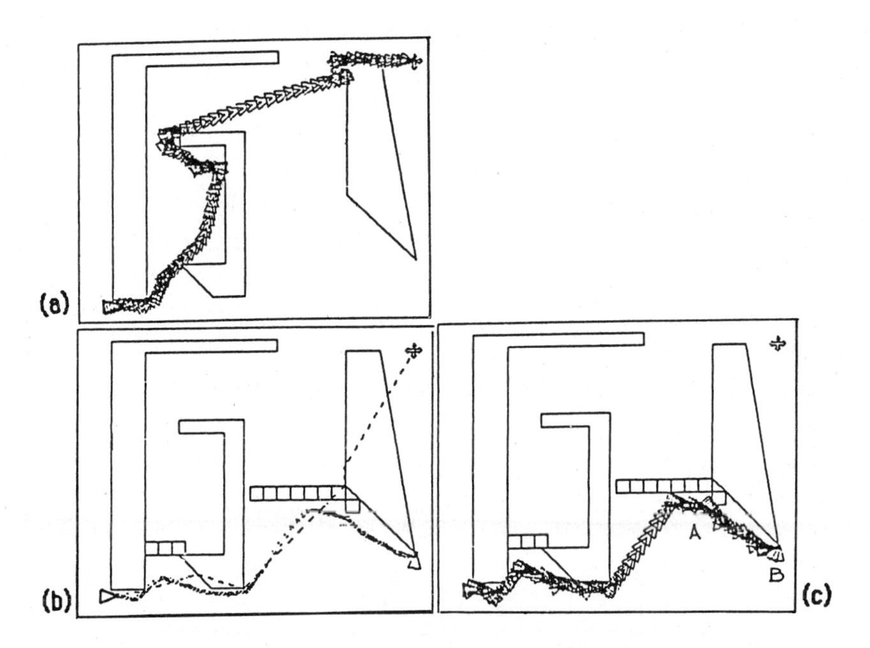

Figure A-2. IMAS operation in Polygonal World (set 2)

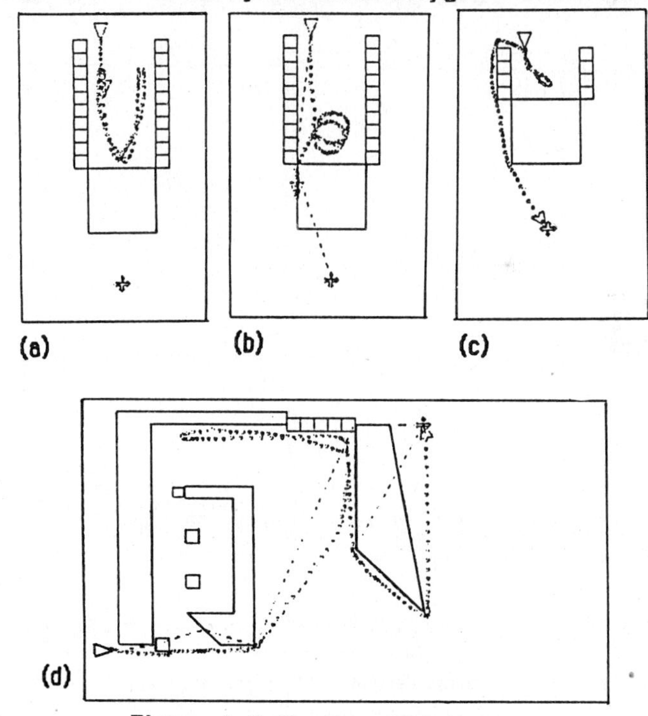

Figure A-3. Dealing with traps

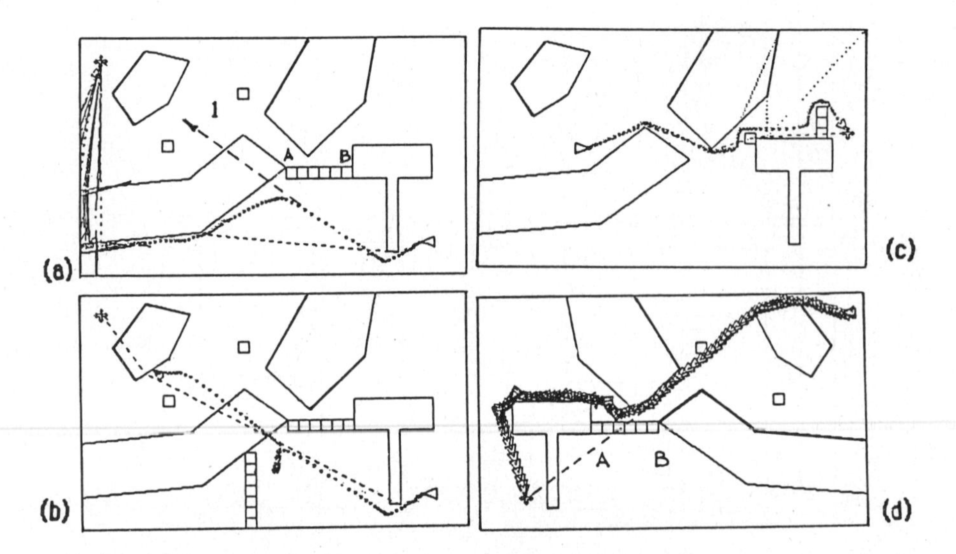

Figure A-4. IMAS operation in Polygonal World (set 3)

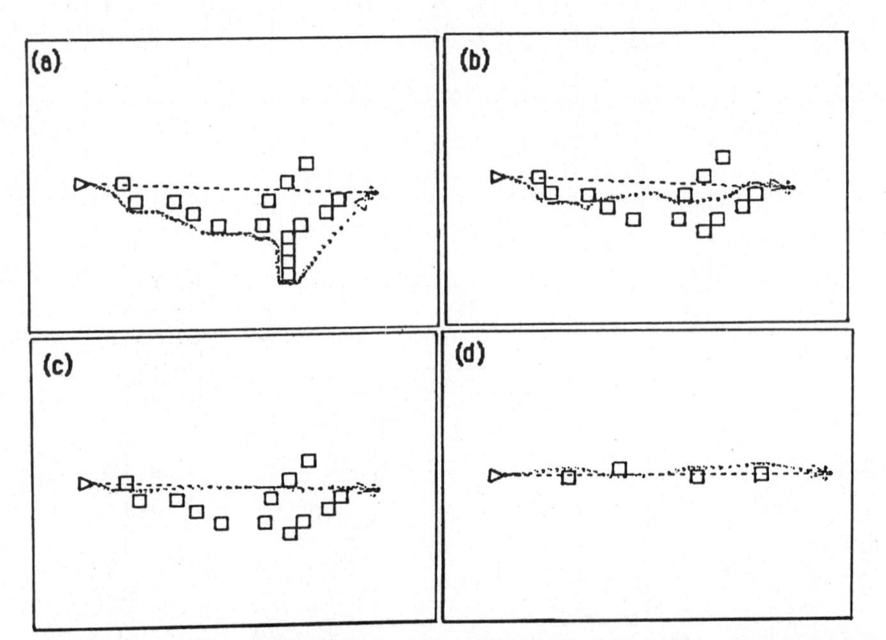

Figure A-5. Simulation of "Pessimistic" and "Optimistic" Strategies

KNOWLEDGE-BASED PILOT AIDS: A CASE STUDY IN MISSION PLANNING

Stephen E. Cross, Robert B. Bahnij, and Douglas O. Norman
Artificial Intelligence Laboratory
Air Force Institute of Technology
Wright-Patterson AFB, OH 45433-6583

1. Introduction.

The published projections of the Air Force Inspection and Safety Center, Mishap Forecast, indicated a total of 62 United States Air Force aircraft would be destroyed in 1985 [1]. 'Pilot error' was projected to be a contributary cause in 40 (65%) of the losses. 28 (70%) of the 40 pilot error related accidents would involve Fighter/Attack aircraft. The following scenario illustrates the problem.

> A four-ship flight of F-16's was on a low level attack training mission to destroy an enemy airfield. The ingress portion was flown at low level in mountainous terrain. The weather deteriorated causing the navigation route to be modified. Assuming that the flight leader did not feel the weather warranted a route abort, the flight proceeded with an alternate route. Typically, under these circumstances, the pilot selects alternate turnpoints and enters them into the navigation computer. Shortly after the flight had turned towards the new turnpoint, the lead aircraft impacted the side of a mountain. No ejection was attempted.

The lead pilot's attention was focused on the task of selecting new turnpoints and entering them into the navigation computer at a time when maximum attention was required outside the cockpit. Apparently, this pilot died because his attention was focused on this task while neglecting other, more important tasks. An important aspect of mission planning is the generation of contingency plans which can be recalled in the air. The following quote from Flying Safety Magazine

---

The views expressed in this paper are those of the authors and do not reflect the official policy or position of the Department of Defense or the United States Government.

succinctly reinforces this point.

> Still another aspect of discipline is planning and personal preparation - to include whatever information is required to handle the mission, plus that required to handle contingencies. The more complex the mission, the more important is thoroughness - avoiding the temptation to cut corners. What's ideal is mental rehearsal - going over every detail in your head ahead of time. Equally important is playing the "what if" game to handle emergencies. Map discipline and route familiarity are critical in the low level arena [2:16].

In this paper, we discuss how the use of artificial intelligence (AI) techniques applied to tactical mission planning systems can help the pilot safely perform time-critical, cognitive processing tasks. Past applications of automation have added new tasks for the pilot to perform (often resulting in task saturation) or have attempted to remove tasks from the pilot (possibly resulting in degraded awareness of the overall mission). We argue that automation must be applied in the appropriate amounts and in the appropriate areas to increase the pilot's performance capabilities. Automation should not arbitrarily remove tasks from the pilot. Rather, automation must be applied in ways that transform the cognitive processing required to perform certain tasks.

The pre-planned, high-speed, low-altitude, surface attack, tactical mission offers a rich domain in which to investigate the design and utility of knowledge-based systems for use in mission planning and onboard pilot aiding. We have chosen this domain for several reasons: (1) there is a crucial need in the US Air Force, as well as NATO forces, to improve the situation awareness of tactical pilots and (2) there are domain experts who are presently studying artificial intelligence at the Air Force Institute of Technology, Wright-Patterson AFB, Ohio. The application area will also be of interest to the German military. The primary mission of the German Luftwaffe F-104 fleet is low-level strike. The primary mission of the German Marine F-104 fleet is low altitude ship reconnaissance and surveillance of the North Sea. German fighter forces have experienced numerous low-level related accidents. In this paper, we discuss prototype systems that have been built and evaluated. Research problems and proposed solutions are discussed.

2. Expectations of Automation.

The first military aircraft had only a few controls and displays such as a throttle, a stick and rudder for flight path control, and a string attached to the fuselage for attitude and yaw display. The pilot's focus was outside the cockpit, largely concentrated on the mission he was trying to perform. As aircraft became more complex, more controls and displays were added. In general, each new control and display was added independently requiring the pilot to act as the integrator. Currently, an F-15 pilot must integrate and assimilate information from over 300 controls and displays [3:38].

The purpose of automation in modern fighter aircraft is to enable the pilot and the entire friendly force to attain the following three objectives.

1. Succeed, or win, in a broad conflict. The pilot interprets this as mission accomplishment (that is, placing his bombs on the assigned target at the assigned time).

2. Help the pilot employ the aircraft and weapon systems to their maximum capability.

3. Ensure force survival. The pilot strives to ensure his own safety, and that of friendly forces. When this safety cannot be ensured, most missions will be aborted, except under extreme circumstances.

A 1982 report on automation in combat aircraft [3] presented four expectations of automation. First, automation is expected to reduce excessive pilot workload. Presently pilots are inundated with disparate data that must be fused and integrated into knowledge which can then be used for decision making. Second, automation is expected to reduce the occurrence of pilot error. Pilots make mistakes when they become task saturated or their attention becomes channeled to a particular task. For example, during a dive bomb delivery pass, the pilot's attention may become fixed on the target, causing him to pull out of the dive too late, and resulting in ground impact. This is called target fixation. Pilots also become distracted, bored, or complacent. Third, overall system performance is expected to improve. For instance, integrated fire and flight control systems may enable new weapon release points. Fourth, automation is expected to yield new capabilities, such as autonomous electronic warfare or terrain following and terrain avoidance (TF/TA) systems. The report concludes the following.

1. Technological advances in the past two decades have made possible the development of complex and more competent aircraft that can successfully operate in more demanding arenas (such as close to the ground, at night, or underneath the weather, and at faster speeds). They can also perform more complex missions, such as simultaneously attacking multiple targets.

2. At the same time, and perhaps as a consequence of these technological advances, the environment in which aircraft must fly and fight has become less tolerant of pilot error. Threats from the air and ground are faster moving, harder to detect, and more difficult to defend against. Not only are the aircraft vulnerable to enemy attack, but given the loss of some onboard systems, they can also be misidentified and shot down accidentally by friendly surface-to-air missiles. [Because of these factors, the number of tasks as well as the cognitive processing for each task have increased].

3. The only stable element over the years is the human operator. The pilot is limited in his ability to assimilate information and perform tasks. He may not be able to cope with the increased workload required to operate today's faster, more highly mechanized and more complex aircraft. Limitations in human capabilities are difficult to overcome, and as yet have not been completely described. The proper use of automation in aircraft could help to overcome these limitations.

4. If automation is truly to improve mission performance, aircraft designers must carefully consider where automation would best serve the pilot's needs. They must examine not only the technology, but the human factors issues involved. This will require an understanding of how pilots process and assimilate information and how they think about their tasks, as well as an understanding of the performance characteristics of the controls and displays through whichthe pilot and the automated system will interact.

An examination of the capabilities of the human problem solver in a real-time environment will illustrate how best to apply automation in ways to improve pilot problem solving capabilities.

3. Pilot As Problem Solver.

People are required to fly aircraft because of their adaptability and excellent pattern recognition and perceptual abilities. These traits allow flexibility unmatched in any conceivable automated system. Pilots have limitations. Pilot reaction times (both physiological and cognitive), limit time critical, inflight problem solving. Because of these limitations, pilots are trained to initially react to emergency, abnormal, and time critical situations by performing memorized "bold face" procedures. The procedures are learned by studying aircraft flight manuals, using checklists, practicing in simulators and training flights, and also from ground -based mission planning. The mission planning process (considering contingencies, asking 'what-if' questions) is a crucial part of preparing a pilot to fly a mission. When inflight novel problem solving is required, the pilot risks task saturation or channelized attention, both of which degrade performance.

Rasmussen [4] provides an illuminating description of man's problem solving behavior. He describes three levels: the skill-based level, the rule-based level, and the knowledge-based level. At the skill-based level his behavior represents "sensory-motor performance during acts or activities which, following a statement of intention, take place without conscious control as smooth, automated, highly integrated patterns of behavior." [4:258]. At the rule-based level, man performs a function when a triggering pattern matches incoming data. The function and situation may be learned from experience and is goal-oriented. The performance is within the context of a domain model that may or may not accurately describe the present domain state. At the knowledge level, goals are explicitly formulated, based on an analysis of the environment and the overall aims of the person. An important aspect of the knowledge-based level is the maintenance of a domain model. This type of knowledge is paramount for asking 'what-if' questions.

Similar observations are made by Santilli, a human factors

researcher at the USAF School of Aerospace Medicine. Santilli describes the three levels as sensation, perception, and decision steps [5]. Decision making is done at a conscious level (involving inductive and deductive reasoning, asking what if questions, etc.) and at a preconscious level where predefined rules are activated.

A typical tactical mission scenario illustrates these levels. This example will be used throughout the remaining sections of the paper. Consider a fighter pilot flying towards his primary target. At some point a validated low fuel warning light, the "bingo" fuel warning light[1], might come on. Amount of fuel is a constraint used in mission planning. The pilot tries to select navigation leg segments that do not violate this or other constraints. This pilot either did a poor job planning or mission events caused the plan to be modified. Based on the mission goals, the pilot must decide to proceed to the primary target, fly towards a secondary target, or return to the base (RTB). If he decides to proceed to the primary target, he may initiate problem solving that determines how he can eventually return to the base. Some options he may consider are not using full afterburner during ingress, target area maneuvering, or egress or planning for air refueling during RTB. These options are context dependent. The problem solving is at the knowledge-based level. The recognition of the bingo light was accomplished at the rule-based level. Perhaps the specific mission plans said to do explicit things if the bingo light came on. Thus, the pilot did not have to contemplate his actions, he simply performed the function associated with the memorized pattern. At the skill-based level, the pilot controlled his trajectory as a result of the decision.

Now suppose the pilot is confident the onboard computer is capable of handling the replanning associated with this task. The fuel low level light may take on a new meaning, namely, "we have a problem and the computer will determine a solution while I do something more important." The pilot was not left out of the loop, nor did he delete the task. Rather he has transferred his knowledge-

---

[1] "Bingo" normally means the "point of no return" fuel amount. To ensure the pilot will have enough fuel to safely return home, he typically adds a 500 pound margin to the bingo fuel amount and enters this new total fuel amount into the aircraft fuel warning system. When this value is reached, the pilot is warned by the flashing "Bingo" light which is mounted on the Heads Up Display (HUD).

based level cognitive processing to the computer and participated in the task accomplishment at the lower cost rule-based level. In essence, the pilot becomes a system manager of certain tasks, a view recently advocated by Garvey [6].

A major claim in this paper is that inflight, time critical human problem solving must be accomplished at the least costly cognitive level. It seems plausible from the above discussion that the cost associated with cognitive processing increases from the skill-based level to the knowledge based level. Thus, automation should help the pilot by allowing him to accomplish the cognitive processing of a task at a lower cost cognitive level. This can be accomplished by providing aids that help generate rules or provide knowledge-based level assistance in the air. Both options support increased pilot capabilities, not by removing tasks from the pilot, but by decreasing the cognitive processing associated with them.

4. A Framework for Assessing Automation.

In this section, we propose a qualitative model that will be useful for understanding the integration of automation in man-machine systems. Our purpose is not to provide exact mathematical descriptions, but rather to provide a framework in which knowledge-based aids can be discussed.

Humans have finite capabilities. They are limited by the amount of information they can process and the amount of time required to process that information. We define attention span as a measure of these limitations. Attention span is a function of psychological and physiological factors. For instance, a pilot's cognitive capabilities may be degraded in a high-g turn or under other stressful situations.

Each task the pilot must perform will be described by cognitive attention (CA) for the task. CA is a measure of the time and information required to perform a task. The measure is dependent on mission factors. For instance, the cognitive attention for a terrain clearance task increases significantly the lower the aircraft flies. Cognitive attention includes the skill-based, rule-based, and knowledge-based problem solving methods required to perform the task. Each problem solving level for an individual task has unique time and information processing requirements. In general, the following constraints hold:

$$CA_S < CA_R < CA_K$$

where

CA_S = the cognitive attention required for task accomplishment at the skill-based level,

CA_R = the cognitive attention required for task accomplishment at the rule-based level, and

CA_K = the cognitive attention required for task accomplishment at the knowledge-based level.

Assume the pilot has n tasks to perform, each with a unique CA. An upper bound on the total required CA is the summation of the individual CA's over the n tasks. An important constraint is that the total required CA cannot exceed the attention span. When this constraint is violated, the pilot is task saturated. Experienced pilots are adept at prioritizing the tasks they must perform, accomplishing tasks at the least costly cognitive level, and deferring or deleting tasks that might cause task saturation. For instance, in low level flight, it seems plausible that terrain clearance tasks (basic aircraft control, vector control, altitude control) have a higher priority than mission related tasks (navigation, threat assessment). Experienced pilots are more likely to recognize when the total required CA exceeds their attention span. When this occurs, lower priority tasks may be deleted or deferred. These observations seem consistent with the experiments conducted in emergency situations with German pilots by Johannsen and Rouse [7]. They observed that planning was directly affected by both the "urgency of time and the occurrence of unanticipated events" [7:278].

With these thoughts in mind, we now define situation awareness. Situation awareness is the management process that controls the sorting and sequencing of mission tasks. It involves both the time to perform a task, information required to do the task, and information about task priority. In some respects, the situation awareness task is analogous to a scheduler in a real-time, event driven operating system. But it is much more complex, because it includes the definition and maintenance of a domain model of the mission environment within which decisions about task priority and deferment can be made. Situation awareness is itself a task. Bahnij provides a detailed presentation of situation awareness [8].

Maintaining situation awareness is absolutely essential in the low altitude, surface attack arena. Domain factors such as increased velocity and low altitude dramatically increase the cognitive attention required to perform terrain clearance tasks. Automation must be applied in such a way as to reduce the pilot's cognitive load. In the past, controls and displays were added independently requiring the pilot to serve as the integrator. The effect has been to increase the number of tasks to be performed. Since each task has an associated cognitive attention, the unfortunate result is the pilot may become task saturated.

Automation can help the pilot in two principal ways. First, automation can provide an environment conducive to rapid ground-based generation and simulation of mission plans. The effect is that the pilot programs himself with rule-based knowledge that can be accessed during flight. Second, automation can help by planning and simulating plans at the knowledge-based level onboard the aircraft in anticipation of pilot-directed or situation-expected recommendations to the pilot. The effect is that the computer assumes the knowledge-based level cognitive processing, allowing the pilot to process the task at a least costly level. At each level, the amount of knowledge and decision aiding assigned to the computer are increased. Artificial intelligence techniques are exploited.

5. Artificial Intelligence.

Two popular definitions of artificial intelligence are:

"the study of ideas that allow computers to make people seem more intelligent" [9] and

"the study of ideas that allow computers to do the things that make people seem intelligent" [10].

The first definition suggests a mission planning environment that allows the pilot to utilize his limited time more effectively and consider more options. The pilot can increase his situation awareness by concentrating on high level mission objectives and considering contingencies. Also, the pilot can program the onboard flight computer to perform certain tasks, much like he would brief a crewmember. The later definition implies that the computer has the capability to perform tasks involving decision aiding because it understands the goals and plans of the pilot.

Problems, like mission planning, can be represented in a tree where the parent node represents the initial world state and the leaves of the tree represent possible world states. Operators can be designed to transform one state into another (node transition). Goals can be viewed as desired world states. Plans are sequences of operators that transform the initial state into a desired goal state. Knowledge-based search techniques are used to find these sequences of operators.

There are a body of programming methods that allow the representation of and reasoning with knowledge. These are called knowledge-based programming (or AI) techniques. An AI technique is any computer programming method that has these characteristics:

(1) exploits explicit representations of knowledge about the problem domain,
(2) the knowledge is understood by the people providing it,
(3) the knowledge representations are easily modified,
(4) the knowledge is useful in situations that are incomplete or uncertain, and
(5) the knowledge partitions the problem.

A key lesson learned in artificial intelligence is that large amounts of domain dependent knowledge are required to limit search. The types of problems which require AI techniques are exponentially hard. This means that the search space grows exponentially as more potential decision points are added. Knowledge decreases, but does not bound, the required search. Another key factor of these techniques is that the knowledge is understood by people who provide it. That is, the same knowledge is used by a domain expert, thus the reasoning behind answers generated by AI techniques are transparent to the user. AI techniques are typically used in applications where the problem is manageably hard and not well understood. Programs are built incrementally, thus it is necessary to be able to quickly modify the program. AI techniques also help solve problems where incoming data is uncertain or incomplete. Lastly, AI techniques are helpful for partitioning problems (implementing "divide and conquer" strategies). Commonly used AI techniques include frame-based systems [11] and rule-based systems [12].

STRIPS was an early planning system in the robotics domain [13]. It employed means end analysis and suffered from combinatorial explosion on simple problems. Hierarchical planning approaches, such as NOAH [14] and MOLGEN [15], limited the search for operators by planning at abstract levels, and filling in details as required. MOLGEN applied different problem solving strategies and posted constraints to deeper, more detailed levels. When constraints could not be satisfied, it applied meta-knowledge to relax constraints. Another popular approach to planning employs scripts [16]. Scripts are precompiled plans. A planner with a library of scripts can splice an appropriate script into a larger plan. Wilensky [17] combines script-based planning, hierarchical planning, and the simulation of plans in an approach to planning and understanding. Our approach to level four automation is based on Wilensky's planning theory.

6. Levels of Automation.

We propose the following automation levels which are similar to those proposed by Hammer and Rouse [18]. At the lowest level are computational aids that allow routine, labor intensive functions such as fuel used or time of flight to be determined automatically.

At a slightly higher level, these functions incorporate knowledge useful for planning and performance tasks. For instance, if during mission planning a fuel constraint is violated, the computer identifies all functions that are affected and affords the pilot an easy way to improve his plan.

Such capabilities allow automation to be integrated at a higher level, to include pilot-directed inflight automation of system monitoring and decision aiding tasks. The pilot can train his computer in flight simulators to perform specific tasks in certain situations. For instance, if the pilot is flying toward a target and knows he is getting low on fuel (confirmed by a fuel low level warning light), he prefers to concentrate on the mission objective, getting his bombs on target. He can do this more effectively if he knows the computer is working the problem of how to get back to the home base. The computer can solve this problem without pilot initiation. By the time the pilot needs a plan for getting home, the computer can make recommendations. The pilot understands the computer's capabilities because he has programmed and trained with it.

At the highest level, the computer initiates problem solving because it understands the goals and associated plans of the pilot and is ready with plausible solutions when the pilot needs help. Because the pilot and computer have trained together, the pilot expects the computer to begin problem solving when the situation dictates it. In the previous example, the pilot would not have to tell the computer to figure out a return home plan - he would assume that the computer was doing so. At this level, the low fuel warning light would take on a new meaning. The pilot's tasks would be to verify the low fuel state and receive the warning light as a message from the computer that it is solving the problem.[2] Again, the proposed taxonomy is along the lines of increased computer cognitive responsibilities with the goal of allowing the pilot to concentrate on high level mission goals. In the remaining parts of this section we discuss prototype systems that have been built to implement the four elements of this taxonomy.

Before discussing specific examples, it is worthwhile to briefly discuss the mission planning process. Again we firmly believe that a priori considerations of what may happen in the air are crucial to formulating the rules upon which critical decisions will later be based. Thus, it is important to give pilots the capability to study their mission, to ask "what if" questions, and to generate alternate plans before they fly. Unfortunately, this is not presently the case. The pilot has about 30 minutes to prepare a

---

[2] Although not addressed in this paper, an important aspect of this level of automation is maintaining a model of the aircraft systems. A diagnosis function would verify the low level warning light was due to flight related factors rather than abnormal conditions (for example, a fuel leak).

detailed mission plan. He must use a ruler, dividers, and a pencil to collect mission relevant coordinates (for example, turnpoints) and develop a plan. He must then estimate or compute with a hand held calculator or microcomputer, the length of the legs, the time to target, and the fuel used. Should constraints such as the assigned time-on-target (TOT) or minimum fuel be violated, he must refine the plan by selecting new coordinates. He must compensate for weather (fog in valley will require him to increase his altitude) and threats (must decide to fly under radar coverage outside lethal range). Because of the time critical nature of mission planning, the pilot cannot replan should these constraints be changed at the last minute (for example, new intelligence information arrives). The plan triggers rule-based behavior in the air. The pilot can perform as a real-time system because he has compiled goal-based knowledge into rules and can organize this knowledge such that it can be retrieved and applied quickly.

6.1. Level 1.

At the first level, many USAF squadrons have introduced 8 bit microprocessors programmed with BASIC computer programs to simplify the generation of mission plans. Some of these systems are integrated with digitizer boards (analog systems that allow a crosshair mouse to select x/y coordinates on a map and an A/D converter which transforms the input into latitude and longitude coordinates). Such systems simplify and reduce the time required to plan a mission. The mission goal is to fly from the home airfield to the assigned target. The pilot chooses a route that minimizes his exposure to threats (by terrain masking) while satisfying the assigned mission objectives of being at the target at the assigned time with a predetermined fuel reserve. The pilot will carry a mission data card to the aircraft and type the coordinates into the navigation computer. As he approaches the turnpoints the aircraft instruments indicate the next turnpoint.

Consider the problems associated with modifying this plan. Suppose that a threat is discovered along the proposed navigation route as the pilot is preparing to leave the squadron building. It takes several minutes to modify the leg to avoid the threat. There is little opportunity for pilot compensation. He is forced into the position of adapting his plan in real-time when he flies near the threat. He may try to fly lower or faster. The obvious disadvantage is that he must concentrate on one specific task (surviving the threat) and this takes his attention away from other mission goals.

6.2. Level 2.

At the second level, we describe a system [8] recently developed by Bahnij, an operational fighter instructor pilot with previous assignments in the F-4, F-104, and most recently the F-16. The goal of the system was to dramatically improve current mission planning. The key to the system is that knowledge critical to mission planning is embedded in the Level 1 functions. The system is centered on a knowledge-based map shown in Fig. 1. The map is constructed from Defense Mapping Agency (DMA) data and is similar in appearance to operational maps. The pilot uses an "electronic grease pencil" (that is, a computer mouse) to select coordinates, specify legs, modify legs or turn points, or perform many other functions. Several important capabilities are presented. When the pilot defines a new leg, all mission critical parameters (fuel used to point, distance flown, time flown, leg distance, leg heading) are computed and displayed. Additionally, continuous straight line distance, time, and fuel required to fly from the present proposed position directly to the target are computed and displayed. This information is used heuristically by he pilot to evaluate the overall "goodness" of his evolving plan. An illustration is shown in Fig. 2. Should a constraint be violated, the pilot can easily change the shape of a route leg or change the location of a turnpoint. All mission parameters are recomputed. Should a threat appear as in Fig. 3, the computer determines what portion of a leg is vulnerable and at what altitude (terrain may mask the aircraft), and suggests avoidance options. The computer may suggest that the aircraft increase speed, decrease altitude, or modify its route. The pilot can select a three dimensional view of a proposed flight path. This is shown in Fig. 4. The profile view increases situation awareness by allowing the pilot to visualize a three dimensional view of his environment prior to flying. Visualization is a well accepted means for improving performance. For instance, Olympic athletes utilized these techniques beginning with the 1972 Munich games.

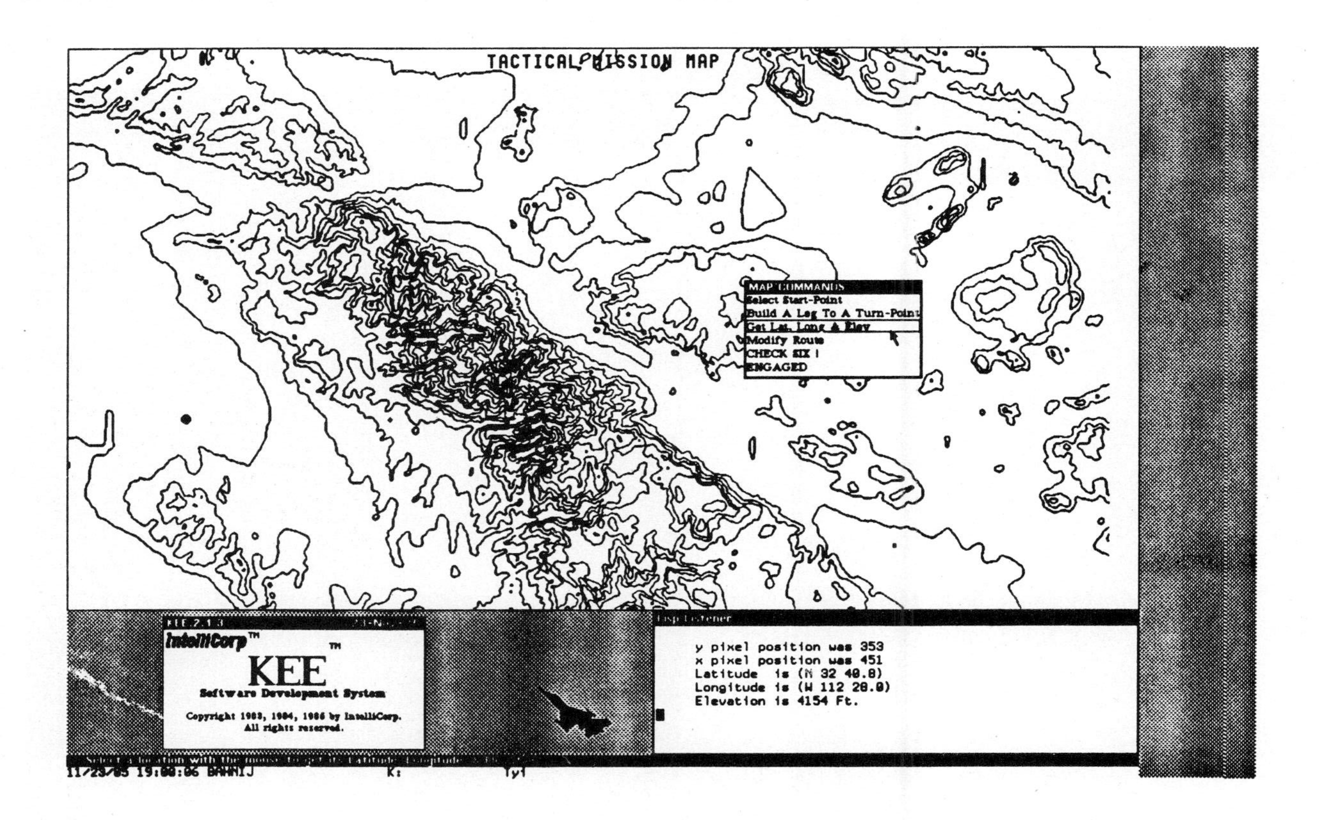

Fig. 1: Knowledge-Based Map

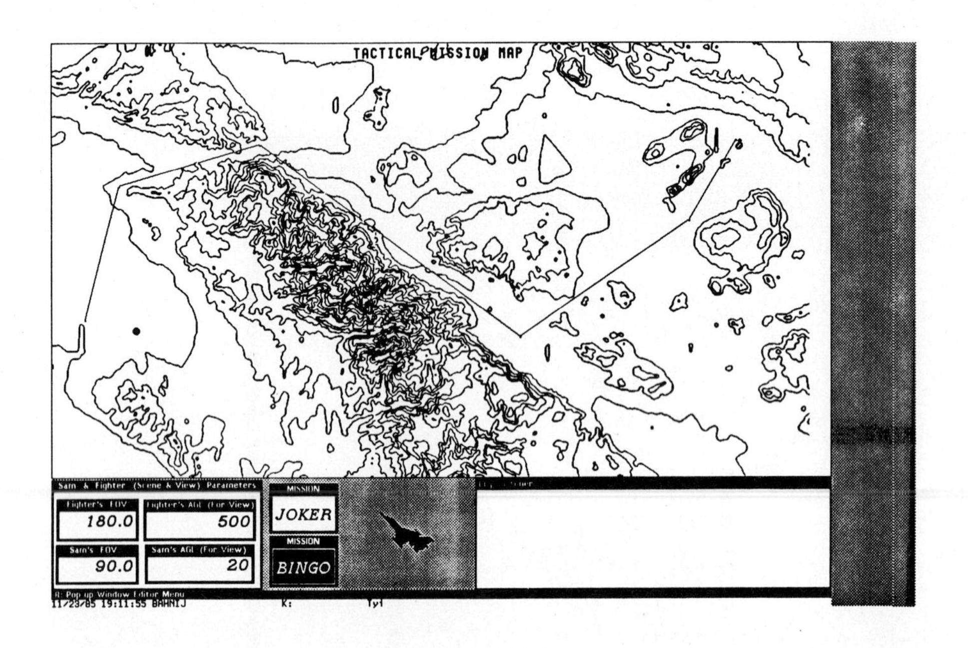

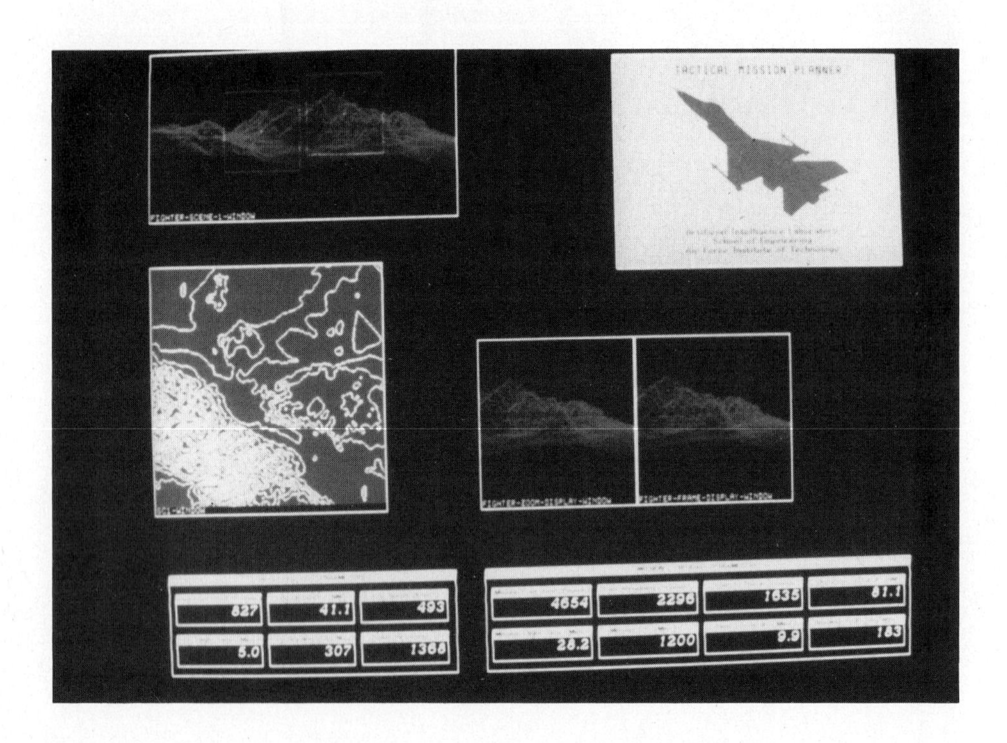

Fig. 2: Knowledge-Based Mission Planning System

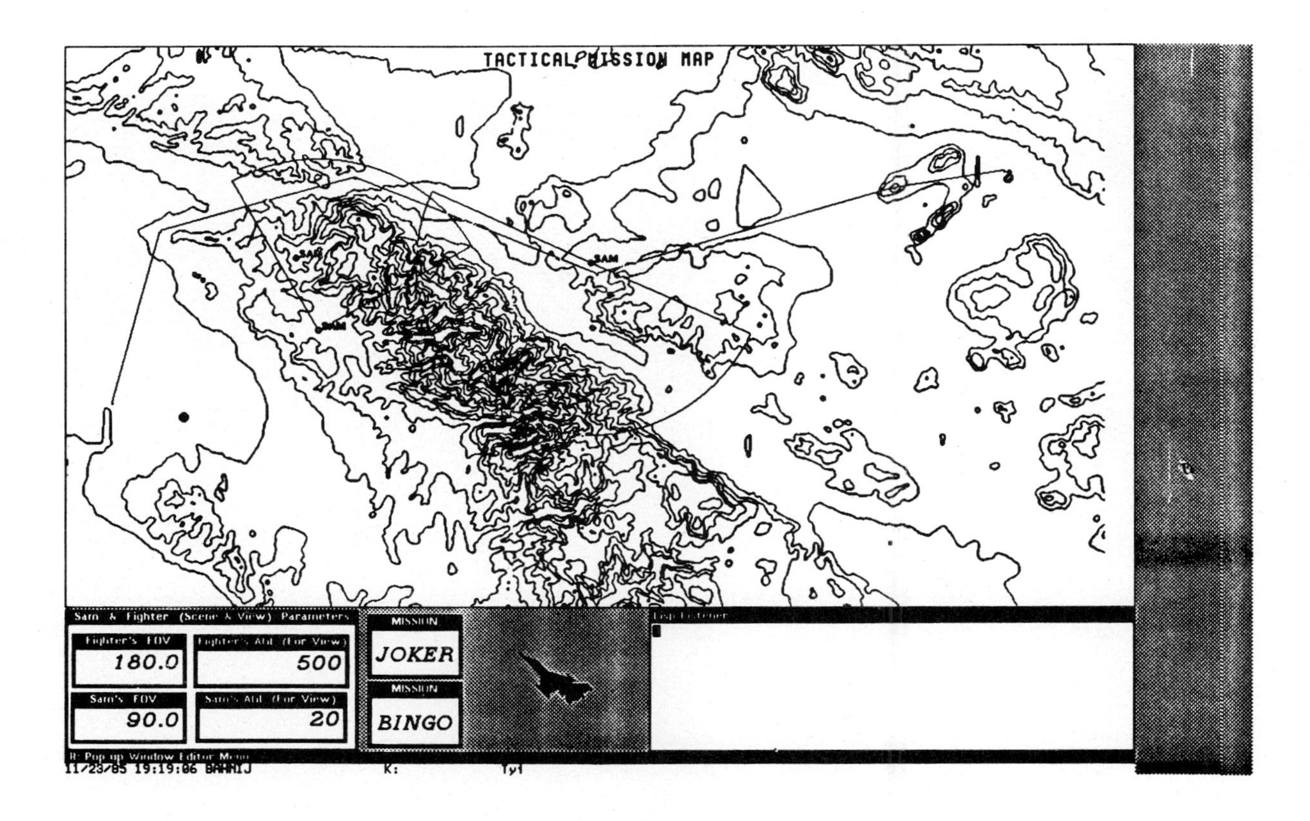

Fig. 3: Threat Depicted on Knowledge-Based Map

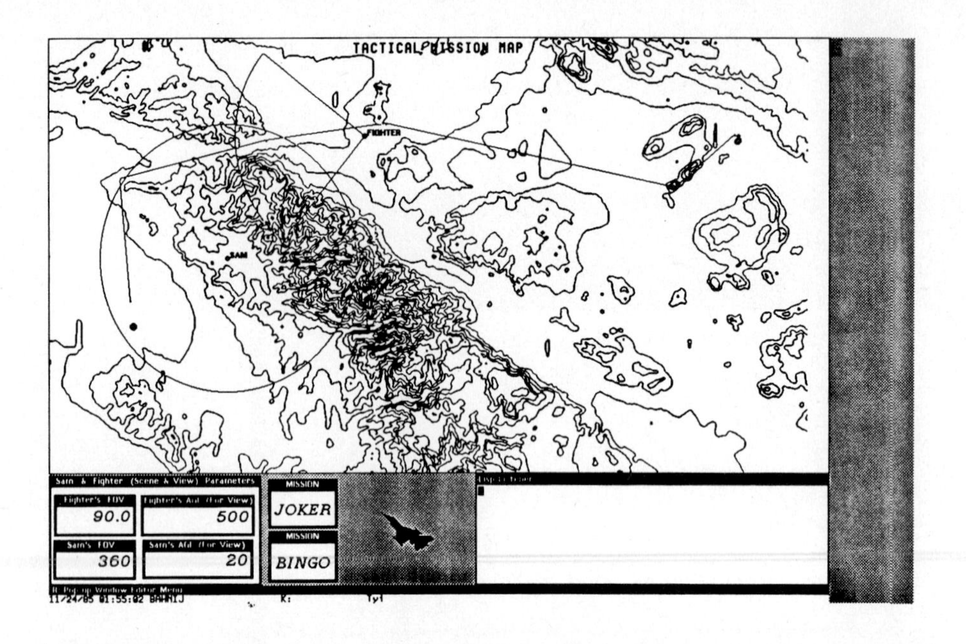

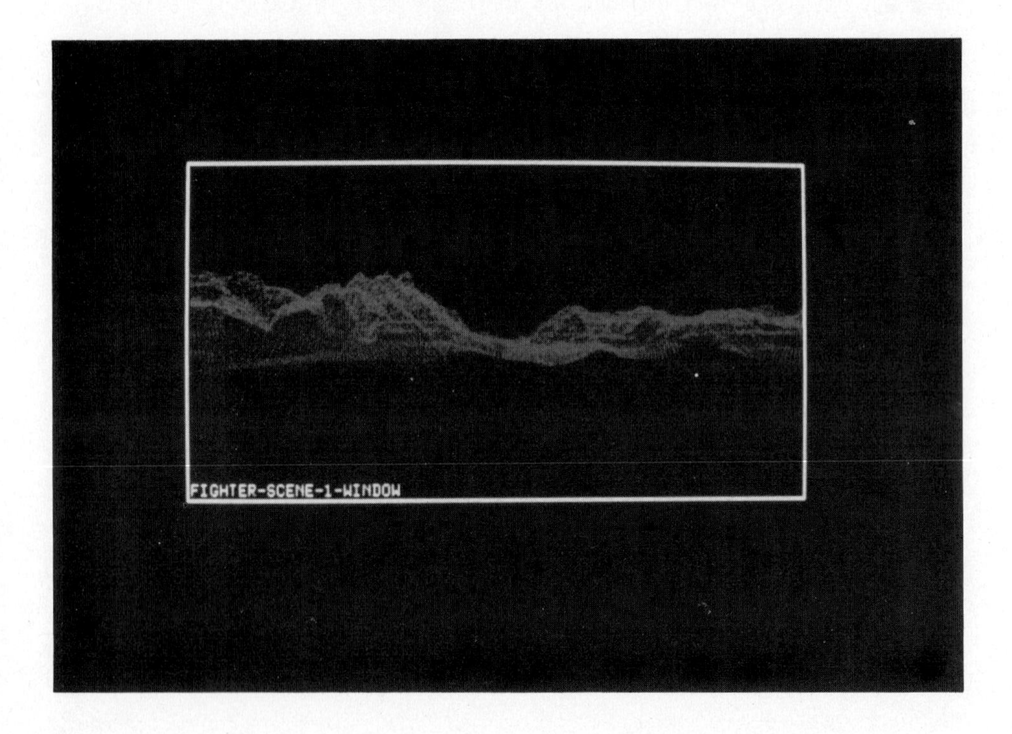

Fig. 4: Terrain Profile Along Pilot-Specified Navigation Leg

The system has several advantages. First, plans can be created quickly and modified easily. Perhaps a more important advantage is that a pilot's total attention is focused upon the map, the mission environment, during the entire planning process. The pilot can hypothesize alternate scenarios and generate contingency plans. Preliminary evaluations by operational pilots have been very encouraging. Although our system is a research prototype, we believe such systems will be included in the Air Force inventory quickly. It provides a knowledge-based language that allows pilots to customize their own systems. It also provides a tool with which we can study the problem solving behavior of pilots.

6.3. Level 3.

The level three capability provides inflight decision aiding for pilot-directed tasks. The intent is to provide the pilot with a computerized 'back-seater.' These tasks may include mission critical planning such as determining a fuel efficient route home, proposing alternate targets and paths around threats, initiating and/or controlling electronic warfare systems, or monitoring safety of flight conditions. The approach is plausible because the pilot understands the capabilities and limitations of the computer, much like he would understand the capabilities and limitations of a human back-seater. This understanding is derived through training. Training is accomplished in simulators and training flights.

The back-seater knows how to perform certain functions, the pilot provides the guidance or sets the thresholds on when these functions are to be performed and how the back-seater is to interface with the pilot. A pilot can program a computer in much the same way. The system described at level two can be augmented with standard software modules that perform various tasks. One such module may include software that monitors fuel flow and determines if sufficient fuel exists to reach the target and return home. If sufficient fuel is not available, the software module would propose plans for returning to a safe location or another acceptable plan. The pilot indicates that he wants the function to be performed at some point in the mission and to be advised at some later point. This serves several purposes. First the pilot understands what the computer will do and when it will do it. Thus, he is fully cognizant of the task, but he does not have to expend costly knowledge-based level processing on the task. He can direct his full attention to achieving the mission objective. The system we propose at this level is an extension to a navigation computer programming system employed by Tornado squadrons at Laurbach, West Germany. Tornado crews carry to the aircraft a cassette tape that is used to program the navigation computer. The same idea is described by Morishige as "giving a pilot a room full of furniture and allowing him to arrange it to suit his tastes" [20:93]. We are extending our level two system so that a pilot could customize his own flight programs.

6.4 Level 4.

A level four capability provides a higher level of automation because now the computer initiates problem solving appropriate for the situation. The intent is to provide each pilot with a surrogate instructor pilot, rather than a back-seater. A crucial assumption at this level is that the computer understand the capabilities and limitations of aircraft systems, including the pilot. Several approaches have been discussed in the literature [17, 21], based on the notion of scripts. A script represents an expected course of action that would be taken by the pilot to achieve some goal. When the computer observes steps in a script it can deduce the goal the pilot is trying to achieve. In the event steps are taken which do not match a given script, the computer can generate plans and test to see if these plans correspond to the pilots actions. In either event, it can be asserted that the computer understands the pilot if it can produce knowledge that predicts (and if requested, explains) the actions the pilot will take to achieve a goal. Thus, the computer can hypothesize the existence of future goals and construct plans that would be acceptable to the pilot. In the next section, we discuss research problems and approaches for building a level 4 system.

7. Research Issues.

In this section, we discuss several ongoing research projects. We are investigating the application of AI planning techniques to improve the level two and three capabilities. At level four, we are conducting research in planning architectures. We are also interested in introducing integrated interfaces into the mission planning environment.

7.1. Planning.

This section discusses the requirements of a planning system for level 4 automation based on the work of Knode [22] and Cross [23] in the domains of system monitoring and air traffic control respectively. The planning system is motivated by the meta-planning ideas advocated by Wilensky. That is, a level 4 system must understand the actions of the pilot. Understanding implies that the computer can deduce the goals the pilot is trying to achieve. It can formulate plans and project these plans into the future. First we discuss some requirements of the planner, then discuss its components, and finally illustrate its capabilities.

7.1.1. Planner Requirements.

7.1.1.1. Autonomous Goal Detection.

The planner must have autonomous goal detection ability, that is, it must not rely on inputs from the pilot to know when and what goal to begin planning for.

7.1.1.2. Understanding.

The planner required to deduce the goal from observed pilot actions. In this view, understanding is an inverse of planning.

7.1.1.3. Conflicting Goals.

The planner must handle conflicting goals. One difficulty of multiple goal satisfaction is that the plan steps for one goal may undo a precondition for another goal. Sussman's HACKER used domain independent knowledge sources, such as the "prerequisite-clobbers-brother-goal," to handle goal conflicts [24]. Cross [23] describes conflicting, overlapping, and cooperative goals and domain dependent methods for their resolution.

7.1.1.4. Simulation.

The planner has to test its proposed plan before presenting it to the pilot or performing the actions specified in it. Testing should involve some method of simulation or projection of the plan into the future and watching to see if the plan will achieve its goal without conflicting with other active goals. One approach using qualitative simulation is described in [25]. If the plan does not work, the planner should learn from this projection how it might improve the plan or suggest a new one. If the new plan would remove another plan's precondition or lacks one of its own, the planner should first try the least expensive way to fix the problem.

The planning task continues after a plan is constructed, simulated, and accepted by the pilot. The planner monitors the consequences of the planned actions. It may have to alter the plan during execution because of new conflicts or unpredicted world events.

7.1.1.5. Model-Based Reasoning.

Cross argued that rule-based architectures were of themselves insufficient to handle the types of reasoning required in level 4 systems [26]. While rules are desirable for execution in time critical environments, rule generation requires a model of the world. In the domain of air traffic control, the computer reasoned that it was proper to descend, rather than climb, a recently inflight refueled aircraft. The computer was able to make this deduction because it wanted to generate a plan that satisfied two interacting goals (aircraft collision avoidance and maintenance of fuel efficient flight paths) and it reasoned from a model based on aircraft equations that the drag of the aircraft was less at a lower altitude. This was significant because there were no existing rules that would lead to this conclusion.

7.2. Level 4 Planner Components.

The planner consists of four components: goal detector, plan proposer, plan projector, and plan executor.

7.2.1 Goal Detector.

The goal detector deduces the goals the pilot is trying to achieve by observing his actions. The goal detector is motivated by Wilensky's Plan Application Mechanism (PAM) program [17] and is similar to Geddes' intent inferencing system [27]. The approach is that the computer has a library of scripts. Each script is a precompiled plan indexed with a goal. The computer can use pattern matching to find scripts that correspond with pilot actions.

7.2.2. Plan Proposer.

The plan proposer receives a goal from the goal detector and searches for relevant plans. For instance, the goal, to "put out the engine fire" would have the plan "Emergency Engine Shutdown" associated with it. Some goals may have more than one applicable plan. Information about the plans is available to facilitate intelligent plan selection.

7.2.3. The Projector.

The projector accepts the proposed plan and projects it into the future beginning with the current world state. The effects and defects of the plan can be studied, offering a chance to improve the plan if necessary. During the projection, detected goals are fed back to the proposer and projector. The projector must have knowledge of the aircraft, its support subsystems, and the aircraft's real-time environment. It has to know if a failure in the simulation is caused by the plan or some other faulty process already in progress. These requirements are recognized to be quite difficult, but regardless of these difficulties, this capability remains a necessity for a flight planner if it is to be truly helpful in the aircraft domain. Therefore, the projector must test the plans by simulation, passing the good ones to the executor and replanning the bad ones.

7.2.4. Executor.

The executor is the fourth component of the planner. It functions as the communicator to the pilot. After the projector passes the approved plan to the executor, the executor communicates the plan to the pilot.

7.2.5. Meta-planning.

The meta-planner recognizes goal interactions as specialized goals and applies knowledge about planning to resolve these goals. Not all planning problems can be solved using the same planning steps; therefore, the nature of the plan interactions should dictate the next planning step.

The sequence of events guided by the meta-planning structure is as follows. The detector notices something of interest and passes the goal to the proposer. The proposer finds the plan for the goal and passes it to the projector. The projector projects the plan into the future and if the detector sees a conflict, it signals a goal conflict. The proposer then looks to see if it has another plan which wouldn't cause the same kind of conflict. If it has one, that plan is tried, if not, the proposer tries to find a plan by reasoning about its model of the domain. Once found, the plan is tested by the projector. If successful, the planning is nearly finished, else the proposer looks to see if the circumstances themselves can be suitably modfied and replanned. If not successful, the goal is abandoned. If successful, the executor communicates the plan to the pilot at the appropriate time.

7.3 Example.

Again we use the low fuel state during target ingress for illustration. At level 4, the pilot's required cognitive attention is dramatically reduced, because he knows the computer will figure out a modified plan. The computer observes that the pilot is performing steps appropriate with mission ingress and deduces that it should begin replanning for egress. Thus the computer will generate a plan that achieves the high level goal of maintaining pilot safety. It first looks in its library of plans. The standard plan might be to modify airspeed and altitude to optimal settings. We quickly see that this would be a foolish thing to do. To implement this plan, the aircraft would have to increase its altitude and decrease its speed. Both actions would cause the aircraft to become vulnerable to enemy threats and to inhibit weapon delivery. Suggesting such an option might cause the pilot to lose confidence in the computer. The computer must project this plan and observe that there is a goal conflict, namely that by increasing the altitude and decreasing the speed, the mission goals cannot be achieved. It looks in the library for other plans. Assuming there are none, it passes the goal conflict as a goal to the meta-planner. The meta-planner applies some specialized knowledge that suggests that one goal be changed. One strategy might be to see if the plan for one goal could be implemented at a later time. Since the mission plan cannot be delayed, the fuel efficiency plan is constrained to be executed after the other plan is achieved. The modified plan is projected. If it works, everything is fine, a plan has been generated for the pilot to use in egress. If the plan does not work, other options can be considered. For instance, standard plans of what to do when fuel is low can be brought to bear such as diverting to alternate airfields, calling for inflight refueling, or using less afterburner during attack or egress.

7.4. Level 2 and 3.

A major problem in the level 2 and 3 systems is constraint interaction. Constraints can be viewed as goals that are not to be achieved. Constraint interactions must presently be resolved by the pilot. For instance, the computer may recommend a route change to avoid a threat. The route change lengthens the overall mission distance and may cause a fuel warning light to come on. Thus, the pilot must consider several replanning options (such as changing velocities, shortening another leg) or relaxing one of the constraints (ignore the fuel warning light). Present thesis projects are investigating the use of constraint satisfaction techniques to handle constraint interactions. The approach is to have the computer do the best it can at resolving the conflict and in the event that a modified plan cannot be found, asking the pilot to relate constraints. The approach appears plausible on the ground. We are not sure if it will be acceptable in the air.

7.5. Integrated Interfaces.

We are investigating methods that will make the interface of the pilot to the mission planning system more natural. Presently, our level 2 system utilizes menus for input and graphics for output. We would like to integrate the mouse and graphics with a natural language system for input. Such a system would allow the pilot to ask questions like:

1. How far is it from <click> to <click>? or

2. Can I complete the mission if I move the IP point to <click>?

The clicks indicate mouse selected points on the mission map. Our initial studies will use typed input, but we will integrate speech recognition systems as the technology matures.

8. A Major Limitation.

A recent thesis by Norman [28] demonstrates that less than hoped for processing time improvements are gained by parallel processing in flight planning problems. The reasons seem obvious in retrospect. A recent discussion on the arpanet illustrates the problem. Suppose the Lisp expression '(cons A B)' is to be processed on a parallel machine. What is the potential decrease in processing time? Ignoring the time to process the cons, the time decrease is at most 50%, assuming that A and B are assigned to separate processors and the time to process each is equal. The processing required of A and B can itself be assigned to other processors. The net result is a decrease in processing time at an exponential increase in memory (or processor requirements). To achieve a three order of magnitude decrease in processing time, 1000 processors would be required. This figure is an upper bound, because it assumes subproblem independence and no overhead.

Norman developed a novel processor assignment algorithm, the parallel paths method, and a simulation language in which plans could be expressed and translated into data flow networks. The data flow networks could be examined and processor assigned. Thus for any given planning problem, the simulation could determine the maximum decrease in processing time for an arbitrary number of processors. The results are quite disturbing. For a typical cross country flight planning problem, the decrease in time is approximately 150%. The improvements over a sequential planner for a cross country type flight planning problem are shown in Fig. 5.

The results suggest some hard upper-bounds on performance increases for a given problem solution. Further, given the nondeterministic nature of search (even with knowledge), an answer cannot be guaranteed in some finite period of time. Thus, using a rich, general reasoning paradigm may be impossible in real-time. This implies a slight alteration in the approach to airborne systems at level 4.

As with humans, proper 'what-if' games must be played prior to the mission. The system examines its model generating these scenarios, both interactively and autonomously, and formulates appropriate "bold face" solutions which <u>can</u> run in real-time., and use a sensitivity analysis to determine which situations deserve solution.

It may be possible to fly with a model which is constructed at a high abstract level (i.e., small enough not to be computationally taxing). If a situation arose where scripts or plans could be instantiated, and no "bold face" procedures were developed during the system's 'what-if' session; the high abstraction model might be used to present a picture to the pilot of the nature of the problem. This approach uses the human to do the thing that humans do best: reason under uncertainty.

9. Related Research Programs.

9.1. Defense Advanced Research Projects Agency (DARPA).

DARPA announced the Strategic Computing Program in 1983. The program has as a goal to produce major research breakthroughs in artificial intelligence, computer architecture, and microelectronics and to demonstrate these research results in military relevant systems. The three demonstration areas are battle management, autonomous weapons, and a pilot associate.

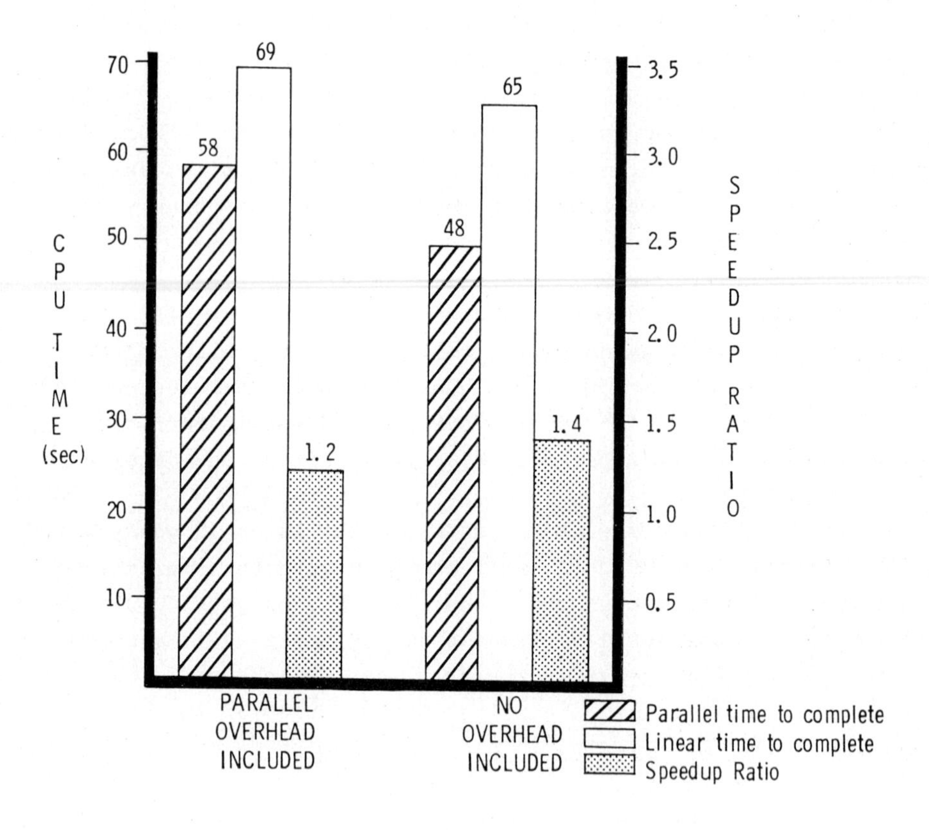

Fig. 5: Improvements From Parallel Processing on a Representative Planning Problem

The pilot associate [29] has five components: mission planner, tactics planner, system monitor, system diagnosis, and pilot-vehicle interface. Proposals are presently being reviewed to choose a contractor for research and development of the pilot associate. The intent is to demonstrate a real-time capability in a flight simulator by 1990.

9.2. Air Force Laboratories.

The Air Force efforts in artificial intelligence as applied to pilot aids are conducted at Wright-Patterson AFB in the Avionics Laboratory, the Flight Dynamics Laboratory, and the Armstrong Aerospace Medical Research Laboratory. The Avionics Laboratory has funded research in emergency checklist procedures. The Flight Dynamics Laboratory conducts research in integrated fire/flight control systems and automated flight control system diagnosis and self-repair. The Armstrong Aerospace Medical Research Laboratory conducts in-house research in speech recognition, cockpit design, and general man-machine research. Of particular interest is funded research in dynamic task allocation [30]. In addition, research at the Rome Air Development Center, Griffiss AFB, New York is directed at mission planning aids. KNOBS [31] is a frame-based resource allocation planning system that specifies aircraft configurations to be applied against various targets.

9.3 Air Force Institute of Technology (AFIT).

AFIT is an accredited graduate school for U.S. and allied military officers and Department of Defense civilians. A five course graduate sequence in artificial intelligence has been offered for three years. Research is conducted in the Institute's Artificial Intelligence Laboratory. The research is funded by DARPA and the Air Force laboratories. A recent discussion of AFIT's research program is in [32].

9.4. National Aeronautics and Space Administration (NASA).

Related research in commercial transport aircraft is conducted by the NASA Langley Research Center, Langley AFB, Virginia [33] and NASA Ames Research Center, Moffett Field, California [34].

10.0. Conclusion.

Successful applications of automation in combat aircraft are the result of the delicate balance of what and how much to automate. Pilot problem solving capabilities can be enhanced in the real-time context by applying automation in such a way as not to decrease the number of tasks facing the pilot, but in ways that provide knowledge-level assistance on those tasks. Four levels of automation are proposed to help the pilot plan and replan low-level, tactical missions. The systems are designed to improve the pilot's situation awareness by decreasing the cognitive attention for various tasks, thus allowing the consideration of more tasks. Level one and two systems are ground-based planning systems which maximize the time a pilot can study the mission environment and generate contingency plans. Level 3 and 4 systems are proposed aircraft systems that can offload some of the knowledge-based level cognitive processing tasks that now likely contribute to task saturation or channelized attention.

11. Acknowledgments.

The authors express their appreciation for the valuable comments from Brig Gen Robert A. Duffy (USAF, Ret.), President, Charles Stark Draper Laboratory, Inc.; Dr. Tom Garvey, SRI International; Lt Col Ron Morishige, USAF; and Dr. Bill Rouse, Georgia Institute of Technology and Search Technology, Inc. The research is sponsored by the Defense Advanced Research Projects Agency and the Air Force Wright Aeronautical Laboratories.

12. Bibliography.

1. "Mishap Report," Flying Safety Magazine, vol. 41, no. 3, March 1985, pp. 1-3.

2. "Hazards of Low Level Flying - Part V," Flying Safety Magazine,

vol. 38, no. 5, May 1982, pp. 14-16.

3. Automation in Cockpit Aircraft, Air Force Studies Board, National Research Council, 2101 Constitution Ave., N.W., Washington, D.C., 20418, 1982.

4. Rasmussen, J. "Skills, Rules, and Knowledge: Signals, Signs, and Symbols, and Other Distinctions in Human Performance Models," IEEE Transactions on Systems, Man, and Cybernetics, vol. SMC-13, no. 3, May/June 1983, pp. 257-266.

5. Santilli, S. "Information Processing and You," TAC Attack, vol. 25, no. 6, June 1985, pp. 6-11.

6. Garvey, T. "AI and Avionics: Keeping the Pilot in Command," keynote address at the First Annual Aerospace Applications of Artificial Intelligence Conference, Dayton Ohio, September 1985.

7. Johannsen, G. and Rouse, W. "Studies of Planning Behavior of Aircraft Pilots in Normal, Abnormal, and Emergency Situations," IEEE Transactions on Systems, Man, and Cybernetics, vol. SMC-13, no. 3, May/June 1983, pp. 267-278.

8. Bahnij, R. A Fighter Pilot's Intelligent Aide for Tactical Mission Planning, MS Thesis, Air Force Institute of Technology, December 1985.

9. Cross, S. "Overview of Artificial Intelligence," unpublished tutorial notes, 1984.

10. Winston, P. Artificial Intelligence, 2nd ed., Addison Wesley Publishing Co., Reading MA, 1984.

11. Fikes, R. and Kehler, T. "The Role of Frame-based Representation in Reasoning," Communications of the ACM, vol. 28, no. 9, September 1985, pp. 904-920.

12. Hayes-Roth, F. "Rule-based Systems," Communications of the ACM, vol. 28, no. 9, September 1985, pp. 921-932.

13. Nilsson, N. "Hierarchical Robot Planning in an Execution System," SRI-AI Technical Memo No. 76, SRI International, Menlo

Park, CA, April 1973.

14. Sacerdoti, E. "Planning in a Hierarchy of Abstraction Spaces," Artificial Intelligence, vol. 5, 1974, pp. 115-135.

15. Stefik, M. Planning with Constraints, PhD Dissertation, Stanford University, CSL Report No. 784, 1980.

16. Schank, R. and Abelson, R. Scripts Plans Goals and Understanding, Lawrence Erlbaum Associates, Publishers, Hillsdale, NJ, 1977.

17. Wilensky, R. Planning and Understanding, Addison Wesley Publishing Co., Reading MA, 1983.

18. Hammer, J. and Rouse, W. "Design of an Intelligent Computer-Aided Cockpit," Proceedings of the IEEE Systems Man and Cybernetics Society Conference, pp. 449-453, 1982.

19. Carter, B. and Nunley, M. "The Micros are Coming," TAC Attack, vol. 21, no. 10, October 1981, pp. 4-7.

20. Morishige, R. and Retelle, J. "Air Combat and Artificial Intelligence," Air Force Magazine, vol. 68, no. 10, pp. 90-93, October 1985.

21. Schank, R. and Reisbeck, C. Inside Computer Understanding, Lawerence Erlbuam Publishers, Hillsdale, NJ, 1983.

22. Knode, D. An Approach to Planning in the Inflight Emergency Domain, MS Thesis, Air Force Institute of Technology, December 1984.

23. Cross, S. "Model Based Reasoning in Expert Systems: an Application to Enroute Air Traffic Control," Proceedings of the 6th Digital Avionics Conference, pp. 95-101, Baltimore, MD., 4-6 Dec 1984.

24. Sussman, G. A Model of Skill Acquisition, PhD Dissertation, MIT, 1974.

25. Cross, S. "Qualitative Sensitivity Analysis: A New Approach to Expert System Plan Justification," Proceedings of the Canadian

Society for Computational Studies of Intelligence, pp. 138-140, May 18-20, 1984.

26. Cross, S. "Requirements of a Flight Domain Expert System Architecture," Proceedings of the International Joint Conference on Systems Engineering, Dayton, Ohio, pp. 250-255, September 1984.

27. Geddes, N. "Intent Inferencing Using Scripts and Plans," to be published in the Proceedings of the First Annual Aerospace Applications of Artificial Intelligence Conference, Dayton, Ohio, 17-19 September 1985.

28. Norman, D. Reasoning in Real-Time for the Pilot Associate: An Examination of a Model Based Approach to Reasoning in Real-Time For Artificial Intelligence Systems Using a Distributed Architecture, MS Thesis, Air Force Institute of Technology, December 1985.

29. Klass, P. "DARPA Envisions New Generation of Machine Intelligence Technology," Aviation Week and Space Technology, April 22, 1985, pp. 46-84.

32. Morris, N., Rouse, W., and Ward, S. "Information Requirements for Effective Human Decision Making in Dynamic Task Allocation," to appear in the Proceedings of the 1985 International Conference on Systems, Man, and Cybernetics, November 12-15, 1985.

31. Engelman, C., Miller, J., and Scarl, E. KNOBS: An Integrated AI Interactive Planning Architecture, Mitre Corp., Bedford, MA., 1983.

32. Milne, R. and Cross, S. "Artificial Intelligence Research at the Air Force Institute of Technology," AI Magazine, vol. 6, no. 1, pp. 74-76, Spring 1985.

33. Baron, S. and Feehrer, C. "An Analysis of the Application of AI to the Development of Intelligent Aids for Flight Crew Tasks," NASA Contractor Report No. 3944, NASA Langley Research Center, October 1985.

34. Chambers, A. and Nagel, D. "Pilots of the Future: Human or Computer?," Computer, November 1985, pp. 74-87.

13. Vitas

Dr. Stephen E. Cross is an Associate Professor of Electrical and Computer Engineering at the Air Force Institute of Technology (AFIT) and Director of AFIT's Artificial Intelligence Laboratory. Dr. Cross received his PhD from the University of Illinois at Urbana-Champaign in 1983. He is a Major in United States Air Force and a graduate of the Flight Test Engineer Course, United States Air Force Test Pilot School. He is a member of Eta Kappa Nu, Tau Beta Pi, the IEEE Computer Society, and the American Association for Artificial Intelligence. He was program chairman for the First Annual Aerospace Applications of Artificial Intelligence Conference.

Robert B. Bahnij is a distinguished graduate of the Air Force Institute of Technology, receiving the Master of Science in Computer Systems in December 1985. His major area of concentration was artificial intelligence. He is a Major in the United States Air Force. Previous to his AFIT assignment, Mr. Bahnij was a fighter instructor pilot and flight examiner, with over 1600 hours, in the F-16, F-104, and F-4. He is a member of Tau Beta Pi, the IEEE Computer Society, the American Association for Artificial Intelligence, and the Association of Computing Machinery. Mr. Bahnij is presently assigned as an Air Operations Staff Officer, Fighter Division, Air Force Center for Studies and Analyses, Pentagon.

Douglas O. Norman is a distinguished graduate of the Air Force Institute of Technology, receiving the Master of Science in Computer Systems in December 1985. He is a Captain in the United States Air Force. His areas of concentration were artificial intelligence and the theory of computation. Presently, Mr. Norman is assigned as a Program Manager at the Electronic Security Agency, San Antonio, Texas.

DYNAMIC PLANNING AND TIME-CONFLICT RESOLUTION IN AIR TRAFFIC CONTROL

Uwe Völckers
Deutsche Forschungs- und Versuchsanstalt
für Luft- und Raumfahrt (DFVLR)
Institut für Flugführung
Flughafen
D-3300 Braunschweig, Germany

1. Introduction

Planning and control of a safe, regular and efficient flow of air traffic at high density airports is an extremely difficult task for Air Traffic Control (ATC). The increasing number of additional requirements that have to be met by ATC, such as: fuel efficiency, noise abatement procedures, wake vortex separation, capacity usage demands have made this task even more complex and challenging.

In a joint project carried out by DFVLR and BFS (the German Federal Air Navigation Services) the COMPAS-System (Computer Oriented Metering Planning and Advisory System) has been developed and is now under tests at DFVLR.

The operational objectives of the COMPAS-system are (with regard to Frankfurt Airport) to achieve the best possible usage of the available, but limited runway landing capacity, to avoid unnecessary delays and to apply economic approach profiles whenever possible. The planning functions which nowadays are still carried out by human controllers will be performed by a computer. It generates and suggests a comprehensive plan for a best overall arrival sequence and schedule. The execution of this plan however intentionally remains the task of the human controllers. They are therefore provided with all necessary data to control the approaching aircraft.

The project objective is to obtain solutions and experience in the layout and application of computer assisted systems in Air Traffic Control.

The design of a semi-automated subsystem necessitates in particular careful and feasible solutions both for the transfer of human planning and decision making functions to a computer and for the distribution of authority between computer and controller.

2. Arrival Planning in Air Traffic Control

2.1 Human planning in todays system

An important task in Air Traffic Control is to merge several converging streams of aircraft from different approach directions on the runway centerline (Fig. 1). On major, often congested airports this is a challenging and complex task. Although the average arrival

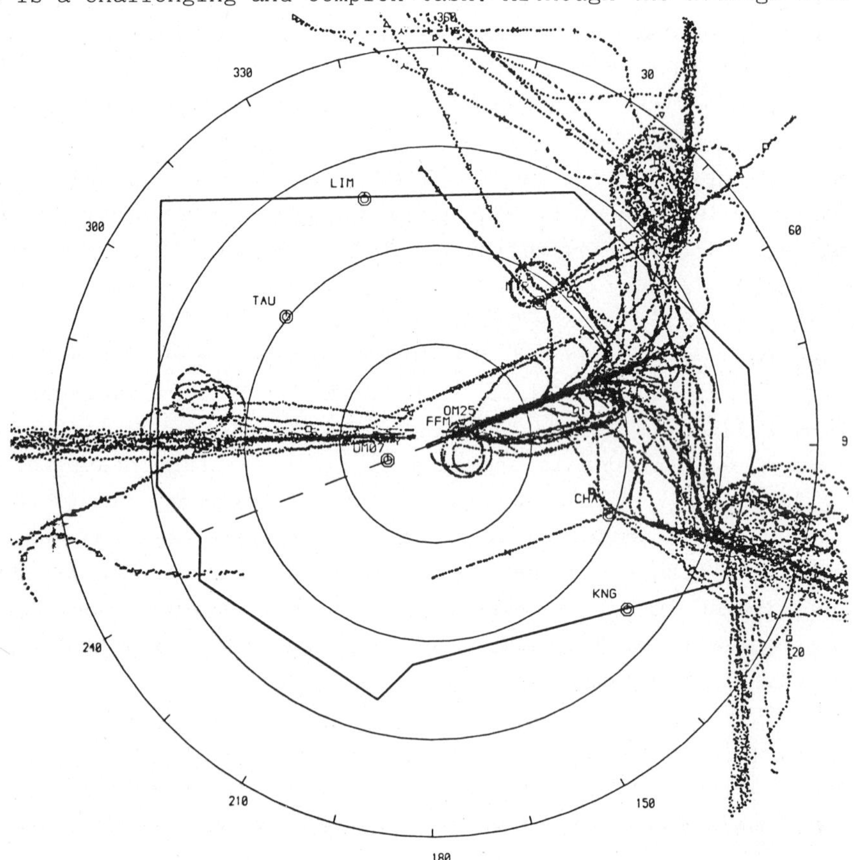

Fig. 1: Radar tracks of arriving aircraft in a 90-min peak period

rate may not exceed the average landing capacity, it cannot be avoided that (despite all long-term flight plan coordination and medium-term flow control) the arrivals are randomly distributed.

This would lead to

- o arrival peaks (with resulting delay),
- o arrival gaps (resulting in unused capacity),
- o uneffective wake turbulence sequencing (capacity reduction) or
- o uneconomic flight profiles

if no appropriate planning and control actions would be taken by air traffic controllers in order to establish a safe, smooth and efficient flow of traffic. The actions should be taken in due time outside the terminal area, to avoid congestion and holding procedures in the narrow terminal area, and to allow the application of economic, idle-descent profiles.

Fig. 2 illustrates in a schematic way the extended approach area of Frankfurt Airport.

A typical approach commences in the vicinity of a so-called "Entry Fix" some 70-100 nm distant from the airport at flight levels between 150 and 280. The different Standard Arrival Routes are converging at three Main Navigation Aids (Clearance Limits), the so-called "Metering-Fixes".

The intermediate approach legs from those three directions finally are merged on the extended runway center line for final approach. Normally the landing sequence should be established at least some 10 nm from the runway threshold. This assumed point is called "Gate".

Today the arrival planning and control process is performed by several control units, some of them assigned to the Area Control Center (units C_i and B_i) others to Arrival Control and Local Control (A_i) (Fig. 2).

The shortcomings of todays situation can be described as follows:

- o The overall-planning task is distributed to several control units.

o Arrival planning is performed "stepwise" from the "outer" units (C_i to B_i) to the "inner" units (A_i).

o Some kind of tactical/ad hoc/local planning prevails in each control unit and must be coordinated with other units.

o The application of one overall planning criterion and concerted control action is very difficult to achieve, because of the very high coordination effort.

o The integration of a variety of data from many different sources has to be performed mainly in the head of the human controllers.

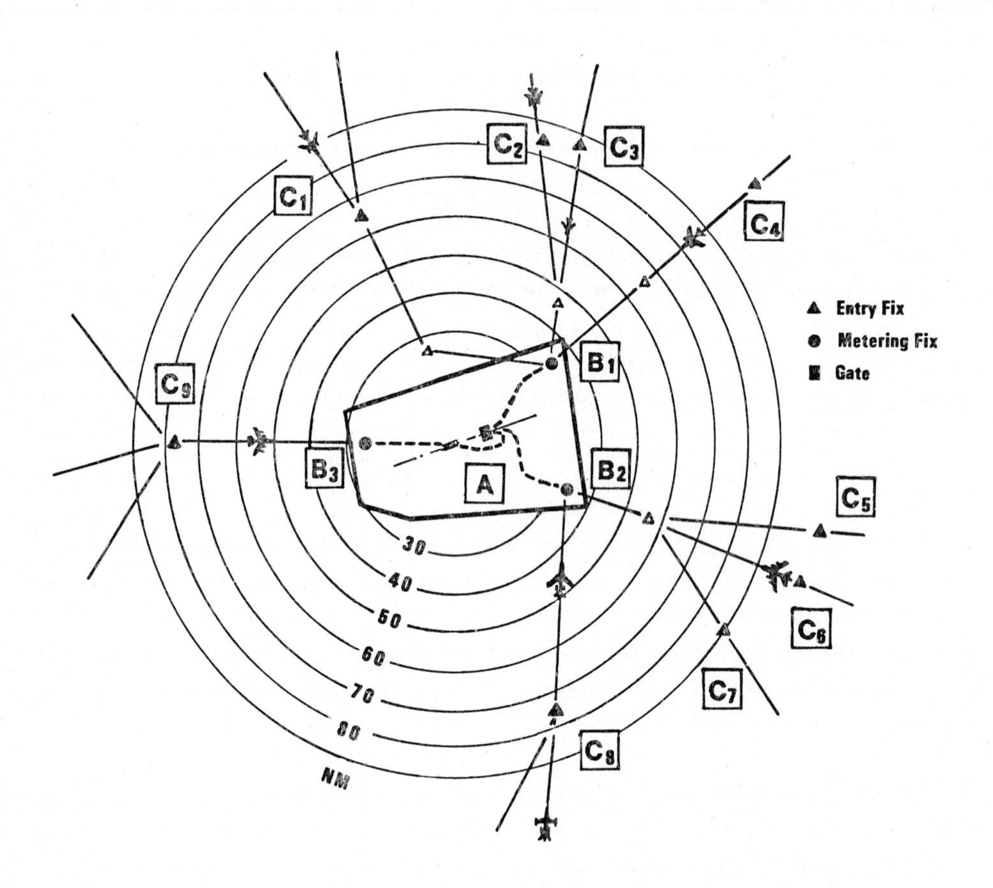

Fig. 2: Structure of the Frankfurt Approach Area

This leads to extremely high work load, and even minor perturbations which could not be matched, may result in a traffic congestion.

Splitting-up and distributing this task to more control units would require even more coordination effort.

Therefore it was envisaged to transfer at least parts of the human planning and control functions to a computer.

2.2 Computer-based arrival planning

Based upon studies at Frankfurt Airport, a concept for a computer based planning system (COMPAS), aiming to assist controllers in the comprehensive planning of arriving aircraft was developed and is now being tested and evaluated by the DFVLR Institute for Flight Guidance in cooperation with BFS - the German ATC-Authority.

The essential design principles of the COMPAS-system can be described as follows:

- The stepwise distributed planning of the human controllers is substituted by one, comprehensive, overall computer planning function.

- The computer planning function anticipates the traffic development for the next 30 minutes and uses one single criterion, common for all units.

- Besides the "usual" data such as radar position information and flight plan data, many other data are included in the computer planning functions, (i.e. traffic load in sub-sectors, aircraft performance and economy, actual airspace structure etc.). The computer integrates these data and generates consolidated planning results.

- Each control unit involved is provided with its specific planning results, necessary to carry out control and to play its part in the overall plan.

- o The controllers stay fully in the loop and keep their executive function. In general the computer generated plan is acceptable to the controllers. However, it is possible for the controllers to interact with the computer in order to modify the plan.

The basic structure of the COMPAS-system is shown in Fig. 3.

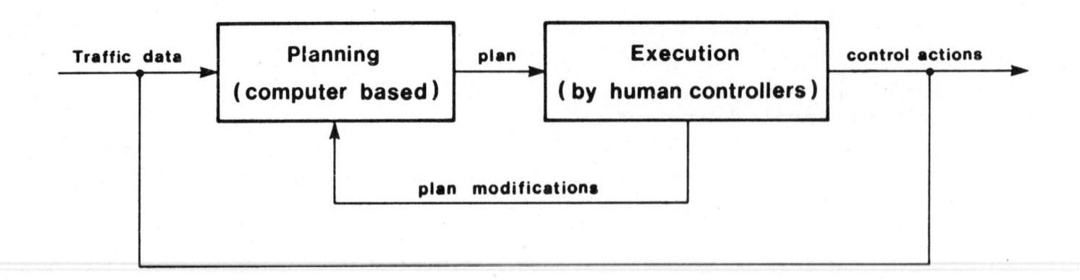

Fig. 3: Planning and execution functions with COMPAS

The operational objectives of the COMPAS-system are with regard to the Frankfurt situation:

- o best usage of runway landing capacity,
- o delay reduction for arrivals,
- o to apply economic descent profiles if possible.

Fig. 4 shows how the COMPAS-system will be integrated into the existing ATC-system.

COMPAS is designed to work in the present ATC-environment. But beyond that actual radar data and flight-plan-data are fed on-line into the COMPAS-DP-system via special interfaces. Taking into consideration many other information (aircraft performance, airspace structure, wind etc.) COMPAS generates a plan and displays it to the controllers. The controller may use these COMPAS-proposals, but is not obliged to use the system. However, in general the results should be so reasonable and convincing, that he easily can adopt these proposals for his control actions. Under normal conditions no controller-computer interaction is required. However interaction is possible, if the controller wants to modify the plan or if it is necessary to cope with unforeseen events.

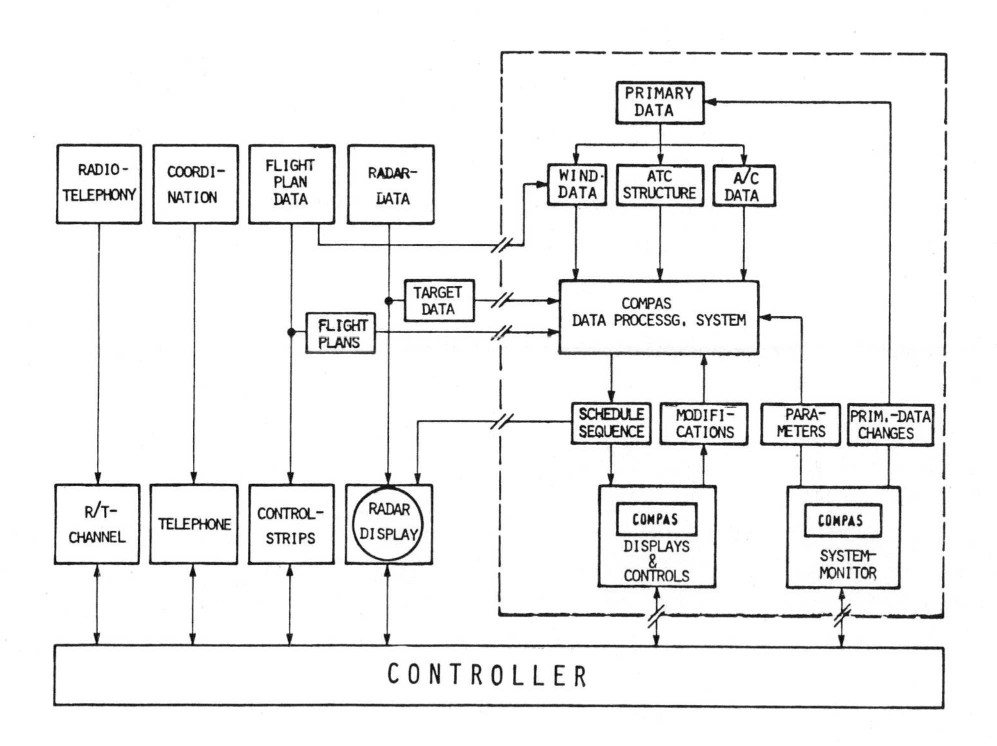

Fig. 4: COMPAS as an additional function in ATC

A successful application of a system like this, where automated and human functions are closely interrelated, strongly depends on the development of feasible solutions for

- o the structure of the planning process,
- o the distribution of authority between controller and computer.

These topics will be discussed in more detail in the following chapters.

3. Dynamic Arrival Planning

3.1 Planning functions

The overall goal of the computer based arrival planning is to generate a plan, giving the "best" sequence and schedule and to provide information how to fulfil this plan. In order to work out a plan dynamically in real time in a real environment, large quantities of data are necessary. They consist of three types of data:

1. Fundamental data and models, which are "static" and do not change during the planning process, i.e.
   - o aircraft performance data;
   - o airspace structure;
   - o approach procedures;
   - o separation values;
   - o wind model.

2. Event-oriented data, which change with "low frequency", i.e.
   - o aircraft entering or leaving the system;
   - o callsign, type of aircraft;
   - o flight plan data (route, way-points, estimated times);
   - o wind data (force, direction).

3. Dynamic data, changing with "high frequency", i.e.
   - o actual flight condition (position, speed, altitude, deceleration rate, descent rate).

Fig. 5 shows the main data-processing functions which are carried out in real-time.

An overview on the various sub-tasks which have to be performed for the planning is listed below. Most of them can be regarded as "auxiliary" functions for the intrinsic planning core: the sequencing and scheduling function, which will be described in more detail.

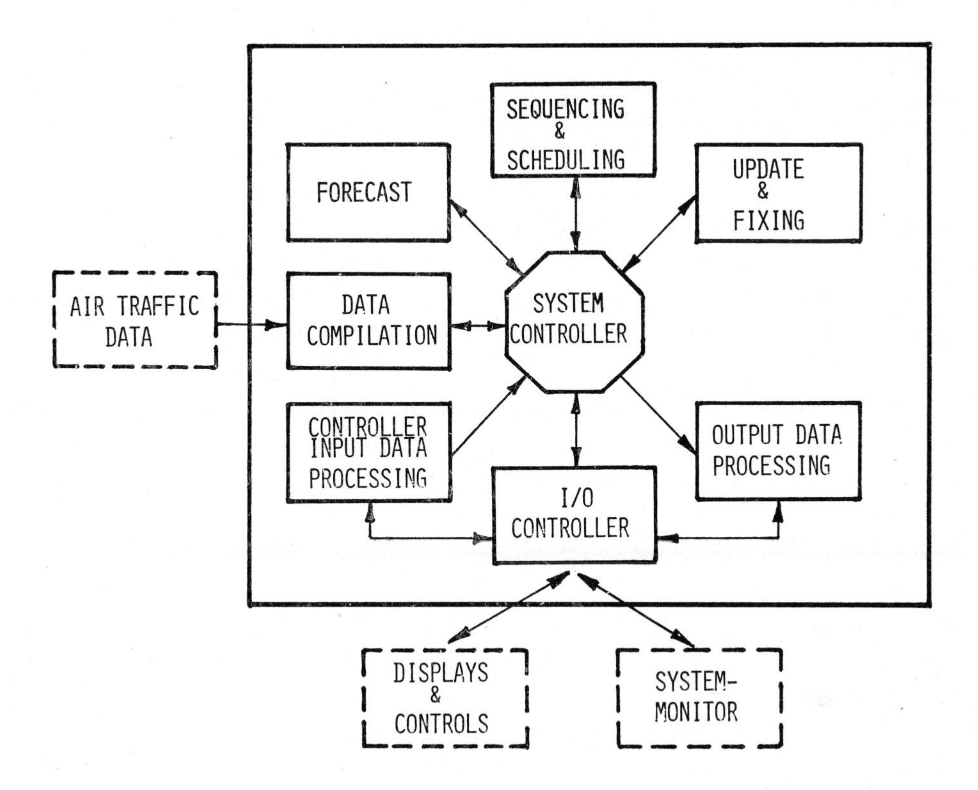

Fig. 5: COMPAS data-processing functions

o <u>System monitoring and control</u>

- Automated DP-controller
- Pre-processing of inputs to change basic-data; sector allocation etc.
- Processing of DP-errors

o <u>Data Compilation</u>

- Radar target recognition, extraction
- Flight plan data selection
- Code/Callsign assignment
- File inauguration and data procurement

o <u>Forecast</u>

- Speed calculation from radar tracking
- Flight path assessment
- Time-to-fly calculation
- Arrival time prediction

o <u>Sequencing and scheduling</u>

- Merging of new arrivals into existing sequence and schedule
- Time-conflict detection
- Time-conflict resolution with "Branch & Bound" algorithm

o <u>Update and Fixing</u>

- Update of Sequence and Schedule
- Assessment of final sequence
- Relief of unnecessary "time-tension"
- "Freezing"of final schedule and sequence

o <u>Input/Output Control</u>

- Controller Computer Interaction Procedures

o <u>Input-Processing</u>

- Processing of controller modifications
- Processing and storage of parameter alterations

o <u>Output-Processing</u>

- Continuous processing and display of results
- Processing and display of data on request

o <u>Basic-Data-Management</u>

- Storage, modification and provision of the valid air space structure, landing-direction, landing-rate, separation, type-specific performance, profile models, wind models etc.

3.2 The Planning Algorithm

The core of the arrival planning function is an algorithm which consists of three major elements:

1. Prediction and initial scheduling.
2. Time-conflict detection.
3. Time-conflict resolution.

The algorithm is activated every time when a new aircraft enters the system.

3.2.1 Prediction and Initial Scheduling

When a new aircraft arrives at an "Entry-Fix", an arrival-time prediction is made for the "GATE". Two arrival times are calculated:

1. The Estimated Time Over Gate (ETOGT) based upon the preferential profile (i.e. idle thrust descent) and all other actual conditions of the flight.

2. The Estimated Earliest Time Over Gate (EETOGT) taking into account all measures to advance the arrival within the performance margins of the aircraft and possible short-cuts of the flight path.

The "earliest" arrival time is used for the inital planning in order to keep the system under "time pressure" and to advance and expedite the traffic flow.

With its EETOGT a new aircraft is inserted into the existing aircraft sequence and schedule for the GATE.

The result is the initial plan, giving a tentative schedule and landing order. Then the time-conflict detection function is called-up.

3.2.2 Time conflict detection

The time-conflict detector is searching the entire landing order for infringements of the minimum permitted time-separation between pairs of two successive aircraft at the GATE.

It uses a data table, the so-called separation matrix (Fig. 6) which gives the minimum permitted time-separation between any combination of leading and trailing aircraft according to their wake-vortex-class.

trailing aircraft / leading aircraft	H HEAVY	M MEDIUM	L LIGHT
H	107	133	160
M	80	80	107
L	80	80	80

Fig. 6: Minimum separation time (sec) between successive aircraft of different wake-vortex-classes corresponding to a landing speed of 135 knots

If there is no time-conflict detected over the GATE, i.e. if the time-separation between any preceeding and following aircraft is equal or greater than the respective minimum separation the planning process is finished.

However, if a time-conflict between two or more aircraft is detected, the conflict-resolution function comes into effect.

3.2.3 Time-conflict Resolution

The time-conflict algorithm works as follows:

1. It consideres the earliest time-conflict in the initial plan. If one of the two aircraft involved in this conflict has its status "frozen" (i.e. its position in the planned sequence over the Gate cannot be changed anymore), then the "non-frozen" aircraft is put behind the "frozen" one, according to the separation matrix in Fig.6. If both aircraft have the "non-frozen" status, there are two possibilities for the sequence:

    - aircraft i behind aircraft j
    - aircraft j behind aircraft i.

    Both possibilities have a given delay time for the postponed aircraft. The time-conflict algorithm now first evaluates the solution with the smaller delay-time to form the revised plan (which is characterized by this delay-time).

2. In this revised plan, in general by postponing an aircraft, new conflicts have been created, which have to be resolved. Again the earliest time-conflict is considered (if there is one remaining). Repeating step 1, this conflict is resolved with the penalty of an additional delay-time. This conflict resolution process continues until a conflict-free revised plan has been found (with a certain total delay-time).

3. In a back-tracking procedure, the algorithm has now to re-check all the solutions of step 1, which had been neglected in the first attempt. For each of these solutions the conflict resolution process of step 2 has to be carried out until:

    - either another conflict-free plan with a smaller total delay-time as that of the previous revised conflict-free plan has been found,

    - or the total delay-time exceeds that of the previous conflict-free plan.

4. This back-tracking procedure is carried out for all neglected conflict solutions of step 1 and 2. The procedure terminates with a conflict-free plan with the minimum possible delay-time.

   The time conflict resolution algorithm is governed by the strategy to minimize the total aircraft delay-time, according to the overall goal of the COMPAS project, to maximize the aircraft throughput. Other strategies are thinkable, e.g. to minimize the total number of time-conflicts to be resolved, thus reducing controller workload.

   The described algorithm is a type of branch & bound-algorithm, which can be visualized as a heuristic-directed search in a tree, using a cost function (the total delay-time). Nodes represent plans and are labeled with the earliest time-conflict to be resolved. The arcs represent the conflict solution procedures "i before j" and "j before i" , as discussed in step 1 above. The tree is developed, using the heuristics "to solve the earliest conflict in the plan first" (Step 1). The total delay-time is summed up along the branches, until a conflict-free plan is reached (Step 2).The value of the cost function for the first conflict-free plan is called "first bound". The back-tracking procedure (Step 3) leads to another branch of the search-tree, which is either closed, when its total delay-time exceeds the value of the first bound, or when a new conflict-free plan is reached with a smaller delay-time (Step 4). The result of this search procedure is a conflict-free plan with minimum delay-time. The procedure is illustrated in Fig.7 .

The algorithm tries to resolve the earliest conflict first, i.e. the conflict between aircraft B an D.

If one of these two aircraft has already been"frozen", in this case aircraft B, the "non-frozen" aircraft D is put behind the leading one. In the next step the algorithm checks if the delay for D has created a new conlict or if other conflicts still exist. Again the earliest conflict is selected, in the example the newly created conflict between D and E. If both sequences D-E or E-D are possible, either D or E may be delayed. The algorithm selects that sequence, where the total delay is minimum. This process is repeated until all time-conflicts have been resolved, giving a first solution for the sequence (1.branch) and the first value of the cost function (1.bound).

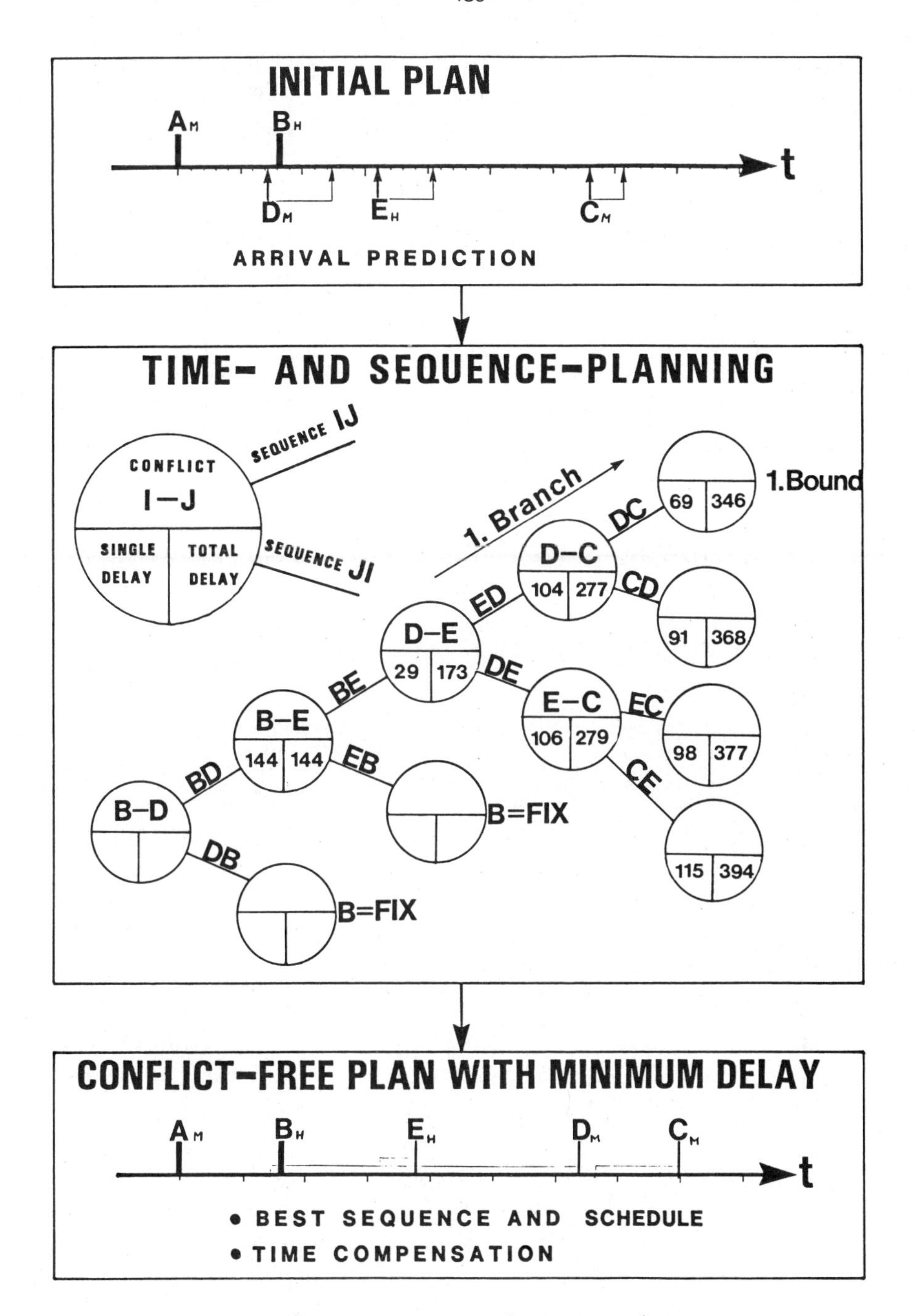

Fig. 7: The COMPAS-planning algorithm (Schematic)

4. Computer-Controller Interface

The layout of the man-machine interface is of very great importance for the practicability and acceptance of a computer assisted function. The guidelines for the layout of the COMPAS-system were:

- o to keep the controller in the loop (i.e. to give him the plan, but to leave the verification of the plan to his experience, skill and flexibility);
- o to display just the necessary data, in a clear and understandable form;
- o to minimize the need for keyboard entries.

These user requirements led to questions and proper solutions for the

- o distribution of authority between controller and computer
- o design of displays and controls and operational procedures.

4.1 Distribution of Authority between Controller and Computer

The requirement was to keep the controller in the loop. This led to a solution where the automated planner permanently carries out the planning functions, with the results (the overall-plan or a sub-plan) being displayed to the respective radar controllers. In the "normal" case, the plan should be reasonable and acceptable and no controller-computer interaction, not even the confirmation of receipt is required.

This means that the control authority fully remains with the controllers. The computer simply takes over the complex planning procedure and makes proposals to the controllers. Assuming that these proposals are compatible with the intentions of the controller and the "behaviour" of the aircraft, the controllers will readily accept the suggested plan, transform it into appropriate control actions, which then are carried out by the aircraft.The traffic situation will then further develop as anticipated by the automated planner. As the controller does not "inform" the computer about his control actions (via data inputs), the computer does have no direct feedback from the controllers, but only monitors and recalculates the development of the traffic situation. Only if modifications are fed in, the computer will react to controller inputs (Fig.8).

The planning function can be classified as "loose, open loop-planning", with - by intention - low accuracy, leaving much responsibility but also flexibility to the human controllers.

Other concepts with a more "tight, closed-loop-planning", are conceivable, however they require even more data, more data-processing capability, more intelligent algorithms and a higher degree of automation.

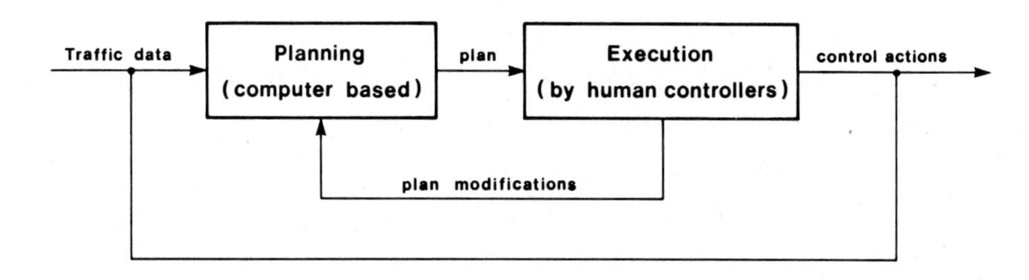

Fig. 8: Distribution of planning and execution functions with COMPAS

In real world operation, however, deviations and disturbances frequently occur and have also to be delt with. So, if the controller has to maintain full control authority, he must be permitted and able to "override" or modify the computer generated plan. In the COMPAS-system this can be done with some function keys. Table 2 shows the operational capabilities of the COMPAS-system without and with controller interaction.

Distributing the authority between controllers and computer causes another problem. In a computer-based system one automated planner generates one overall plan, which then is divided into several sub-plans and distributed to the respective controllers in the different sub-sectors (Fig. 9).

COMPAS-OPERATIONAL CAPABILITIES

CONTINOUSLY/AUTOMATICALLY

- SEQUENCING & SCHEDULING
  (SEQUENCE; ARRIVAL TIMES FOR MF AND GT)
- METERING CONTROL ADVICE
  (SPEED-, DESCENT, FLIGHT-PATH, RECOMMENDATION)
- ADDITIONAL INFORMATION (DISPLAY ON REQUEST)

IF NECESSARY, WITH CONTROLLER INTERACTION

- CHANGE OF NOMINAL SEPARATION
- CHANGE OF LANDING DIRECTION
- CHANGE OF ATC/AIRSPACE STRUCTURE
- CHANGE OF SEQUENCE
- INSERTION OF ARR. INTO SEQUENCE
- EXTRACTION OF ARR. OUT OF SEQUENCE
- EXCEPTIONAL CASES AND PROCEDURES

Table 2: COMPAS-Operational Capabilities

This means the computer has and generates some kind of "master-plan". If this overall-plan is not apparent in the sub-plans, the sub-plans might not be transparent, understandable and acceptable to the controllers. Therefore it is important to provide information on the overall-plan, be it "on-request" or permanently.

Another problem resulting from the distribution of sub-plans is, that plan modifications may originate at different places. This leads to questions of priority, of conflicting interactions and of deterioration of the general goal of the planner, and as well to the stability of the planning process. For the COMPAS experimental system with a limited number of controller working stations satisfying solutions have been worked out. In an operational system application with a great number of controller working stations this problem has to be resolved carefully.

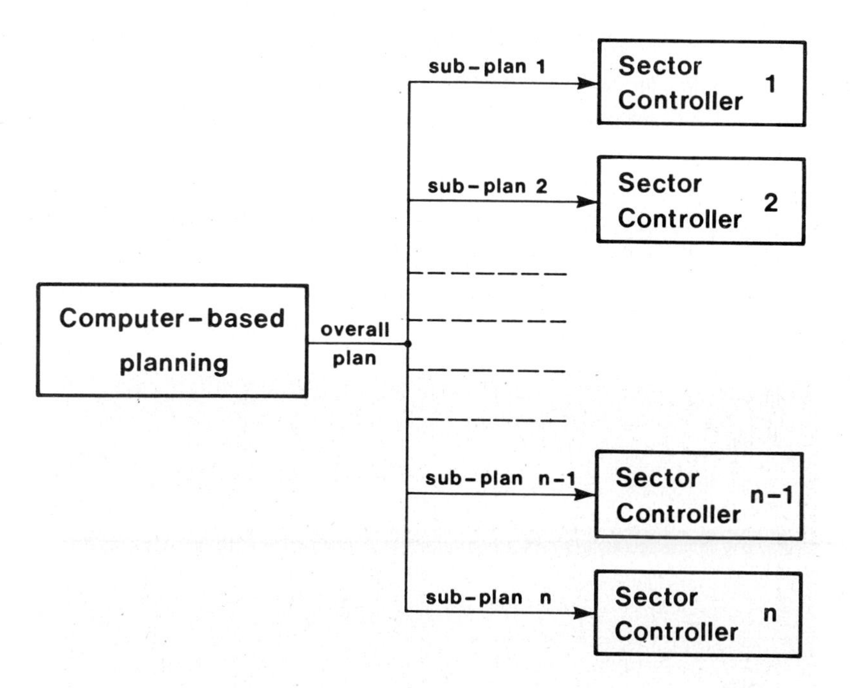

Fig.9: Sub-dividing the overall-plan

4.2 Displays and Controls

As mentioned above the user requirements are:

- to display just the necessary data, in a clear and understandable form,

- to minimize the need for keyboard inputs.

This led to simple, but very clear displays and functions keys.

A coloured display is used. The basic version of the display for the arrival controllers is shown in Fig.10 .The display shows at top right:

- the landing direction in use; (25)

- the airport acceptance rate; (Flow 3.0 means: unrestricted flow, with 3 nm minimum separation when permitted).

The left part of the display shows a time-scale for the next 20 minutes, with the actual time (10.17h) at the bottom. The expected arrivals in this sector are displayed with their call-sign and wake-vortex-class (H). The leading aircraft is at the bottom (according to the typical arrangement of the flight-progress-strips on the strip-holder-board). The small box on the bottom left represents the GATE, giving the indication that, e.g. the JU 358 should be over the GATE right at 1o.17h, followed by the AY 821 about 85 sec later, a.s.o..

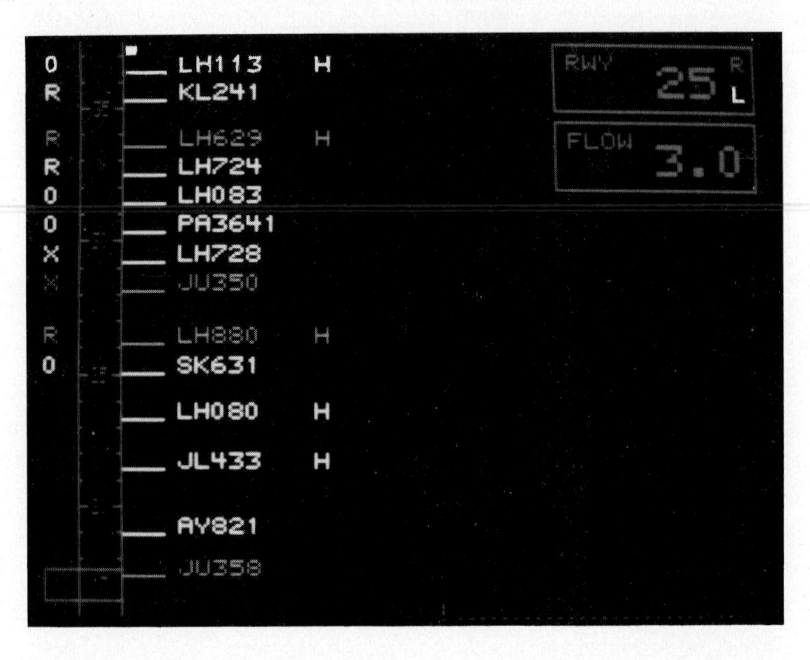

Fig. 10: COMPAS-display of proposed sequence and schedule

The letters left of the time-scale give a rough indication of the suggested control action. Four qualitative suggestions are made to the controllers in order to establish a smooth, dense landing stream:

"X" - expedite (30 sec up to 2 minutes),

"O" - no action ($\pm$ 30 sec),

"R" - reduce (30 sec up to 3 minutes delay),

"H" - hold (more than 3 minutes delay).

As an example: the LH 880 should arrive at the GATE at about 10:26 and it should be reduced. Because the LH880 is a HEAVY-type aircraft, increased wake-turbulence-separation is planned for the succeeding JU350, which has to be expedited in order to catch its landing slot.

There is no proposal for the specific control command. Whether speed control, delay vectors or a combination of both is to be applied, is left to the judgement and experience of the controller, who will consider the entire traffic situation.

The displays for the sector controllers are configured accordingly. However, the bottom box then corresponds to the "Time over the Metering Fix". Displayed is the whole sequence, i.e. the sequence to be merged from all approach directions. According to the colour of the stripholders used in the different approach sectors the labels are presented in the respective colours, giving the controller a clear indication from which direction an aircraft could be expected and giving a hint for what reasons the computer possibly has made a different proposal than the human controller would have done with his limited knowledge of the overall situation.

In case of higher degrees of automation or in case even more sophisticated "intelligent" planning algorithms are applied, the questions of transparency and understanding become even more important, as the controller must be able to fully monitor the automated control process and to take over control at any time, in case of emergency.

In this application of a semi-automated sub-system the solutions provided for transparency and acceptance were worked out in close cooperation with the users.

As mentioned above, the controller is allowed to modify the computer generated plan if he desires or if unforeseen events have to be matched.

Fig.11 shows the small functions-keyboard which is used for controller-computer-interaction. There are 8 (2 spare) function-keys to activate the operational interventions described above. In addition there are "Clear"- and "Execute"-keys and the so-called "-/+"-keys which are used either:

o to move a cursor down or up the time-scale, in order to identify or modify the plan of a specific aircraft, for example for a sequence change;

or

o to increase or decrease parameter values; e.g. if the flow rate shall be changed: after pressing the FLOW-key, first the valid value is displayed on the input-control-line (bottom-right). It then can be increased or decreased with the "-/+"-keys. The desired value is activated with the EXECUTE-key.

All input procedures are performed in this same manner.

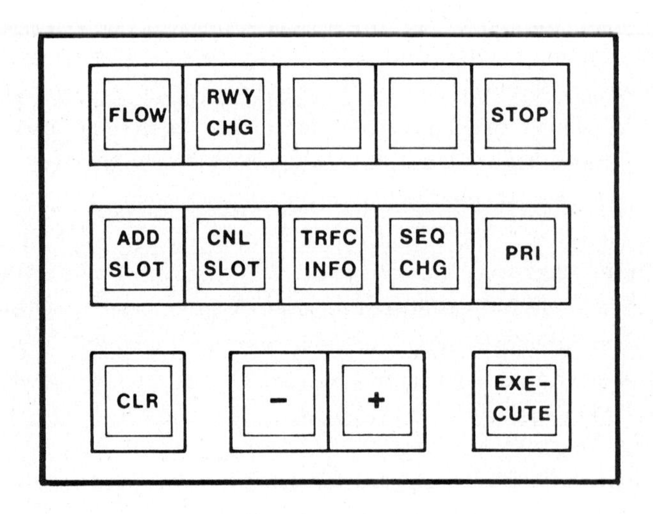

Fig. 11: COMPAS-function keys

Although much more sophisticated displays and controls are conceivable and feasible from an engineering point of view (as much more information is available in the computer) the COMPAS development strictly adhered to the design-requirement of keeping the man-machine-interface as simple and clear as possible.

The first trials with controllers from Frankfurt are very promising with regard to acceptance and operational feasibility.

5. Conclusions

The described computer-based planning system has been developed in all its elements at the DFVLR-Institute for Flight Guidance. It has been tested and evaluated at the institute's air traffic simulation facility, using traffic scenarios of Frankfurt Airport in real-time simulations, with up to 52 aircraft movements simultaneously.

The dynamic planning algorithm as well as the operational concept for computer assistance and the man-machine interface not only proved to be feasible, but were also readily accepted by more than 30 air traffic controllers from the Frankfurt Air Traffic Control Center, who took part in the tests and evaluations.

Thus a first step towards the introduction of intelligent computer assistance for the controllers has been successfully achieved. It is however quite obvious, that this step of transferring human planning and decision making functions to a computer is still limited, with respect to the operational requirements of the user. A next step, which has been started in cooperation with Prof. Wahlster's institute at the University of Saarbrücken, is to implement some human controller heuristics in a rule-based system, which will then be coupled with the described algorithm. Essential for any operational application however are not only appropiate models and suitable computer capabilities, but in particular the careful design of elements and procedures for the man-machine interface.

Outline of a Computational Approach to Meaning and Knowledge Representation Based on the Concept of a Generalized Assignment Statement*

L.A. Zadeh
Computer Science Division
University of California
Berkeley, CA 94720

1. Introduction

The concept of an assignment statement plays a central role in programming languages. Could it play a comparable role in the representation of knowledge expressed in a natural language? In our paper, we generalize the concept of an assignment statement in a way that makes it a convenient point of departure for representing the meaning of propositions in a natural language. Furthermore, it can be shown -- though we do not stress this issue in the present paper -- that the concept of a generalized assignment statement provides an effective computational framework for a system of inference with propositions expressed in a natural language. In some ways, this system is simpler and more direct than predicate-logic-based systems in which it is the concept of a logical form -- rather than a generalized assignment statement -- that plays a central role [7,16,23,24,25,26,28,30,31].

The approach described in the present paper may be viewed as an evolution of our earlier work on test-score semantics and canonical forms [36,38,41]. In test-score semantics, a proposition, *p*, is viewed as a collection of elastic constraints, and its meaning is represented as a procedure which tests, scores, and aggregates the constraints associated with *p*, yielding a vector test score which serves as a measure of compatibility between *p* and what is referred to as an *explanatory database*. The main advantage of test-score semantics over the classical approaches to meaning representation such as truth-conditional semantics, possible-world semantics and model-theoretic semantics [3,8,17,20,21,30,31], lies in its greater expressive power and, in particular, its ability to deal with fuzzy predicates such as *young*, *intelligent*, *near*, etc. [2,5,11,18,22,32,36,44]; fuzzy quantifiers exemplified by *most, several, few, often, usually*, etc. [10,33,40]; predicate modifiers such as *very, more or less, quite, extremely*, etc. [35,44]; and fuzzy truth-values exemplified by *quite true*, *almost true*, and *mostly false* [36].

*Research supported in part by NASA Grant NCC-2-275 and NSF Grant IST-8320416.

The concept of a generalized assignment statement serves to place in a sharper focus the representation of a proposition in a natural language as a collection of elastic constraints. More specifically, in its generic form, the generalized assignment statement may be expressed as

$$X \ isr \ \Omega \,, \tag{1.1}$$

where X is the constrained variable; Ω is the constraining object, usually an n-ary predicate; and *isr* is a copula in which r is a variable which defines the role of Ω in relation to X. The usual values of r are: d, standing for *disjunctive*; c, standing for *conjunctive*; p, standing for *probabilistic*; g, standing for *granular*; and h, standing for *hybrid*. Since in most cases the value of r is d, it is convenient to adopt the convention that *isd* may be written more simply as *is*.

In (1.1), the generalized assignment statement is unconditioned. More generally, the statement may be *conditioned*, in which case it may be expressed as

$$X \ isr1 \ \Omega1 \ \text{if} \ Z \ isr2 \ \Omega2 \,, \tag{1.2}$$

in which Z is a conditioning variable; $\Omega2$ is an object which constrains Z; and $r1$ and $r2$ are variables which define the roles of $\Omega1$ and $\Omega2$ in relation to X and Z, respectively. In general, both X and Z may be vector-valued.

Disjunctive and Conjunctive Constraints[1]

As a simple illustration of a disjunctive constraint, if X is a variable which takes values in a universe of discourse U and Ω is a subset, A, of U, then the generalized assignment statement

$$X \ is \ A \tag{1.3}$$

signifies that the value of X is one of the elements of A. In this sense, A may be interpreted as the *possibility distribution* of X, that is, the set of its possible values [11,36,39].

More concretely, consider the proposition

p: Mary left home sometime between four and five in the afternoon.

In this case, if X is taken to be the time at which Mary left home, the meaning of p may be represented as the generalized assignment statement

$$X \ isd \ [4pm, 5pm] \,, \tag{1.4}$$

[1] The discussion of disjunctive and conjunctive constraints in the present paper is based on earlier discussions in [36,38]. Recent results may be found in [34].

or more simply, as

$$X \ is \ [4pm, 5pm] \ ,$$

in which the interval [4pm, 5pm] plays the role of a unary predicate.

As an illustration of a conjunctive constraint, consider the proposition

p: Mary was at home from four to five in the afternoon.

In this case, if X is taken to be the time at which Mary was at home, the meaning of p may be represented as:

$$X \ isc \ [4pm, 5pm]. \tag{1.5}$$

Note that in this case X takes *all* values in the interval [4pm, 5pm].

The assignment statements (1.4) and (1.5) differ from conventional assignment statements in that the assignment is set-valued rather than point-valued. Furthermore, although the assigned sets are identical in (1.4) and (1.5), they play different roles in relation to X. The possibility that the same constraining object may constrain X in different ways is the principal motivating reason for employing in (1.1) a copula of the form *isr* in which the variable r specifies the role of Ω in relation to X.

In the examples considered so far, the constraint induced by Ω is inelastic in the sense that there are only two possibilities: either the constraint is satisfied or it is not, which is characteristic of constraints associated with assignment statements in programming languages. In the case of natural languages, however, the constraints are usually elastic rather than inelastic, which implies that Ω is a fuzzy predicate. As a simple example, in the case of the proposition

p: Mary is young

the constrained variable, X, is the age of Mary, and the predicate *young* may be interpreted as an elastic constraint on X characterized by the function μ_{young}: [0, 100] $\rightarrow$ [0, 1], which associates which each numerical value, u, of the variable *Age* the degree to which u fits the definition of *young* in the context in which p is asserted. In this sense, $1 - \mu_{young}(u)$ may be interpreted as the degree to which the predicate *young* must be stretched to fit u.

Probabilistic Constraints

As was alluded to already, a proposition p may have different generalized assignment statement representations depending on the intended meaning of p. For example, the proposition

$$p\text{: } Madeleine \text{ is } tall \tag{1.6}$$

may be represented as a disjunctive statement

$$X \text{ is } TALL \ , \tag{1.7}$$

in which $X \triangleq Height(Madeleine)$ and $TALL$ is a unary fuzzy relation which is the denotation of the fuzzy predicate *tall*.[2] The fuzzy relation $TALL$ is characterized by its membership function μ_{TALL}, which associates with each numerical value of height, h, the degree, $\mu_{TALL}(h)$, to which h fits the intended meaning of *tall*. Equivalently, $TALL$ may be interpreted as the possibility distribution, Π_X, of X. In this interpretation, (1.7) may be represented as

$$\Pi_X = TALL \ , \tag{1.8}$$

with the understanding that the possibility that X can take h as a value is given by

$$\pi_X(h) \triangleq Poss\ \{X = h\} = \mu_{TALL}(h) \ , \tag{1.9}$$

where π_X represents the *possibility distribution function* of X.

Alternatively, the proposition *Madeleine is tall* may be interpreted as a characterization of the probability distribution of the variable *Height(Madeleine)*. If this is the intended meaning of (1.6), then the corresponding generalized assignment statement would be probabilistic, i.e.,

$$X \text{ isp } TALL \ , \tag{1.10}$$

in which $r = p$ and $TALL$ is a probability distribution. Thus, if P_X is the probability distribution of X, then (1.10) may be represented as

$$P_X = TALL \ . \tag{1.11}$$

It should be noted that in the absence of a specification of the value of the copula variable r, the proposition

$$p\text{: } Madeleine \text{ is } tall$$

may be interpreted as a possibilistic constraint on $X \triangleq Height\ (Madeleine)$, as in (1.7), or as a probabilistic constraint, as in (1.10). We shall assume that, unless it is specifically stated that the intended interpretation of a proposition, p, is probabilistic or conjunctive, p should be interpreted as a possibilistic, i.e., disjunctive constraint. This understanding reflects the assumption that in natural languages the constraints implicit in propositions are preponderantly possibilistic in nature.

---

[2] Here and in the sequel, denotations of predicates are expressed in uppercase symbols. The symbol $\triangleq$ stands for *is defined to be*.

A related point that should be noted is that in the possibilistic interpretation of (1.6), the value of $\pi_X(h)$ or, equivalently, $\mu_{TALL}(h)$, may be interpreted as the conditional probability of the truth of the proposition *Madeleine is tall* for a given h. In the context of a voting model, this is equivalent to viewing $\mu_{TALL}(h)$ as the proportion of voters who would vote that *Madeleine is tall* given that her height is h [13,14]. Although these interpretations are of help in developing a better understanding of the properties of the membership function, it is simplest to regard $\mu_{TALL}(h)$ as the degree to which h fits the predicate *tall* in a given context, or, equivalently, as $1 - \sigma$, where σ is the degree to which the predicate *tall* must be stretched to fit h.

Granular Constraints

In the case of a granular constraint, the generalized assignment statement assumes the form

$$X \text{ } isg \text{ } G \text{ ,} \tag{1.12}$$

where X is an n-ary variable $X = (X_1, \ldots, X_n)$, and G is a *granular distribution* expressed as

$$G = \{(p_1, G_1), \ldots, (p_k, G_k)\} \text{ ,} \tag{1.13}$$

in which $p_1, \ldots, p_k$ are positive numbers in the interval [0,1] which add up to unity,[3] and the $G_j, j=1,\ldots,k$, are distinct fuzzy subsets of a universe of discourse U.

The generalized assignment statement (1.12) may be interpreted as a summary of n possibilistic assignment statements, each of which involves a component of X, i.e.,

$$X_1 \text{ } is \text{ } G_{j_1} \tag{1.14}$$

$$\ldots$$

$$X_n \text{ } is \text{ } G_{j_k} \text{ ,}$$

in which each G_{j_s}, $s = 1, \ldots, k$, is one of the G_j. In this collection of statements, p_j is the proportion of X's which are G_j.

As an illustration, consider the following proposition

> *p: There are twenty residents in an apartment house; seven are old, five are young and the rest are middle-aged.* (1.15)

In this case, X_i *is the age of ith resident*, $i = 1, \ldots, 20$; $n = 20$; $k = 3$; $G_1 \triangleq OLD$; $G_2 \triangleq YOUNG$; $G_3 \triangleq MIDDLE.AGED$; $p_1 = 7/20$; $p_2 = 5/20$; and $p_3 = 8/20$.

---

[3] A more detailed discussion of the concept of a granular constraint and its role in the Dempster-Shafer theory of evidence may be found in [37].

Hybrid Constraints

A hybrid constraint is associated with a generalized assignment statement of the form

$$X \text{ } ish \text{ } \Omega \text{ ,} \tag{1.16}$$

and may be viewed as the result of combination of two or more generalized assignment statements of different types, e.g.,

$$\frac{\begin{array}{c} X \text{ } isr1 \text{ } \Omega_1 \\ X \text{ } isr2 \text{ } \Omega_2 \end{array}}{X \text{ } ish \text{ } \Omega \text{ .}}$$

An important special case of a hybrid constraint is associated with the concept of a hybrid number [19]. In this case, the constraint on X is characterized by two generalized assignment statements of the form

$$\begin{array}{c} Y \text{ } is \text{ } A \\ Z \text{ } isp \text{ } P \end{array} \tag{1.17}$$

and the relation

$$X = Y + Z \text{ ,}$$

in which A and P are, respectively, possibility and probability distributions, and X is defined to be the sum of Y and Z. In terms of A and P, the constraining object Ω in (1.16) may be viewed equivalently as a *probabilistic set* [15], a random fuzzy set [14], or a fuzzy random variable [27].

2. Meaning Representation

As was stated already, the basic idea underlying test-score semantics is that a proposition in a natural language may be interpreted as a collection of elastic constraints. Thus, by expressing the meaning of a proposition, p, in the form of a generalized assignment statement, we are, in effect, answering two basic questions: (a) What is the constrained variable X in p; and (b) What is the constraint, Ω, to which X is subjected?

In more concrete terms, the process of representing the meaning of a proposition, p, in the form of a generalized assignment statement, X *isr* A, involves three basic steps.[4]

1. Constructing a collection of relations $\{R_1, \ldots, R_k\}$ in terms of which the meaning of p is to be represented. The meaning of each of these relations is assumed to be known, and each relation is assumed to be characterized by its name, the names of its attributes and the domain of each attribute. For our purposes, it is convenient to

[4] For simplicity, our discussion of these steps is limited to the possibilistic case.

refer to the collection $\{R_1, \ldots, R_k\}$ as an *explanatory database* or *ED* for short, and to regard each relation as an elastic constraint on the values of its attributes. It should be noted that the concept of an explanatory database is related, but is not identical, to that of a collection of possible worlds [8,17,29,21,31].

2. Identifying the variable X which is constrained by p and constructing a defining procedure which computes X for a given explanatory database.
3. Constructing a procedure which computes the constraint A as a function of ED.

To illustrate this process, consider the proposition

> *p: Over the past few years Naomi earned far more than all of her close friends put together.*

To represent the meaning of this proposition, assume that the explanatory database consists of the following relations (+ should be read as *and*):

$$\begin{aligned} ED = \; & INCOME\,[Name;\, Amount;\, Year] \; + \\ & FRIEND\,[Name1;\, Name2;\, \mu] \; + \\ & FEW\,[Number;\, \mu] \; + \\ & FAR.MORE\,[Income1;\, Income2;\, \mu]\;. \end{aligned} \tag{2.1}$$

In this database, the relation *INCOME* associates with each $Name_j$, $j = 1,\ldots,n$, $Name_j$'s income in year $Year_i$, $i = 1, 2, 3, \ldots$, counting backward from the present; in *FRIEND*, μ is the degree to which *Name1* is a friend of *Name2*; in *FEW*, μ is the degree to which the value of the attribute *Number* fits the definition of *few*; and in *FAR.MORE*, μ is the degree to which *Income1* is far more than *Income2*.

Next, we have to construct a procedure for computing the constrained variable X. Assume that X is the total income of Naomi over the past few years. Then, the following procedure will compute X.

1. Find Naomi's income, IN_i, in $Year_i$, $i = 1, 2, 3, \ldots$, counting backward from present. In symbols,

$$IN_i = {}_{Amount}\,INCOME[Name = Naomi; Year = Year_i]\;, \tag{2.2}$$

which signifies that *Name* is bound to Naomi, *Year* to $Year_i$, and the resulting relation is projected on the domain of the attribute *Amount*, yielding the value of *Amount* corresponding to the values assigned to the attributes *Name* and *Year*.

2. Test the constraint induced by *FEW*:

$$\mu_i = {}_{\mu}FEW[Year = Year_i]\;, \tag{2.3}$$

which signifies that the variable *Year* is bound to $Year_i$ and the corresponding value of μ is read by projecting on the domain of μ.

3. Compute Naomi's total income, X, during the past few years:

$$X = \Sigma_i \mu_i IN_i, \tag{2.4}$$

in which the μ_i plays the role of weighting coefficients. Thus, we are tacitly assuming that the total income earned by Naomi during a fuzzily specified interval of time is obtained by (a) weighting Naomi's income in year $Year_i$ by the degree to which $Year_i$ satisfies the constraint induced by *FEW*, and (b) summing the weighted incomes.

The last step in the meaning representation process involves the computation of A. In words, A may be expressed as *far more than the combined income of Naomi's close friends over the past few years*. The expression for A is yielded by the following procedure.

1. Compute the total income of each $Name_j$ (other than Naomi) during the past few years:

$$TIName_j = \Sigma_i \mu_i IName_{ji}, \tag{2.5}$$

where $IName_{ji}$ is the income of $Name_j$ in $Year_i$.

2. Find the fuzzy set of close friends of Naomi by intensifying the relation *FRIEND* [35]:

$$CF = CLOSE.FRIEND = {}^2FRIEND, \tag{2.6}$$

which implies that

$$\mu_{CF}(Name_j) = ({}_{\mu}FRIEND[Name = Name_j])^2,$$

where the expression

$${}_{\mu}FRIEND[Name = Name_j]$$

represents $\mu_F(Name_j)$, that is, the grade of membership of $Name_j$ in the set of Naomi's friends.

3. Compute the combined income of Naomi's close friends:

$$CI = \Sigma_j \mu_{CF}(Name_j) TIName_j, \tag{2.7}$$

which implies that in computing the combined income, the total income of $Name_j$ is weighted with the degree to which $Name_j$ is a close friend of Naomi.

4. The desired expression for A is obtained by substituting CI for *Income2* in *FAR.MORE* and projecting the result on *Income1* and μ. Thus

$$A = {}_{\mu,\, Income1}FAR.MORE\ [Income2 = CI]. \tag{2.8}$$

In summary, the meaning of p may be represented as the possibilistic assignment statement (1.3) in which the constrained variable, X, is given by (2.4), and the elastic constraint on X is expressed by (2.8). In essence, the possibilistic assignment statement (1.3) defines the possibility distribution of X given p. What this means is that A, as expressed by (1.8), associates with each numerical value of *Income1*, the possibility that it could be far more than the combined income of Naomi's close friends over the past few years.

The same basic technique may be applied to the representation of the meaning of a wide variety of propositions in a natural language. In the following, we present in a summarized form a few representative examples.

Example 1.

$$p\text{: } Richard\ is\ blond\ . \tag{2.9}$$

In this case

$$p \rightarrow Color(Hair(Richard))\ is\ BLOND\ , \tag{2.10}$$

where $\rightarrow$ stands for *translates into.*

Example 2.

$$p\text{: } Brian\ is\ much\ taller\ than\ Mildred\ . \tag{2.11}$$

Here X is a binary variable (X_1, X_2) whose components are

$$X_1 = Height(Brian)$$

and

$$X_2 = Height(Mildred)\ .$$

The elastic constraint on $X = (X_1, X_2)$ is characterized by the fuzzy relation *MUCH.TALLER*. Thus,

$$p \rightarrow (Height(Brian),\ Height(Mildred))\ \text{is}\ MUCH.TALLER$$

is the possibilistic assignment statement which represents the meaning of (2.11).

Example 3.

$$p\text{: } most\ Swedes\ are\ blond\ . \tag{2.12}$$

In this case, the constrained variable X is the proportion of blond Swedes among the Swedes. More specifically,

$$X = \Sigma\ Count(BLOND/SWEDE)\ , \tag{2.13}$$

where the right-hand member expresses the *relative sigma-count* [40] of blond Swedes among the Swedes. Thus, if the individuals in a sample population in Sweden are labeled *Name1*, ..., *Namen*, then

$$\Sigma\ Count(BLOND/SWEDE) = \frac{\Sigma_i\ \mu_{BLOND}\ (Name_i) \wedge \mu_{SWEDE}\ (Name_i)}{\Sigma_i\ \mu_{SWEDE}\ (Name_i)} \tag{2.14}$$

in which $\mu_{BLOND}(Name_i)$ and $\mu_{SWEDE}(Name_i)$ represent, respectively, the degrees to which $Name_i$, $i = 1, ..., n$, is *blond* and Swedish, and the conjunctive connective $\wedge$ yields the minimum of its arguments.

The elastic constraint on X is characterized by the possibility distribution of the fuzzy quantifier *most*, which is a fuzzy number *MOST*. From (2.13) and (2.14), it follows that the possibilistic assignment statement which represents the meaning of (2.12) may be expressed as

$$p \rightarrow \Sigma\, Count(BLOND/SWEDE) \text{ is } MOST \ , \tag{2.15}$$

in which the constrained variable is given by (2.14).

3. Inference

One of the important advantages of employing the concept of a generalized assignment statement for purposes of meaning representation is that the process of deductive retrieval from a knowledge base is greatly facilitated when the propositions in the knowledge base are represented as generalized assignment statements. This is a direct consequence of the fact that a generalized assignment statement places in evidence the variable which is constrained and the constraint to which it is subjected.

Viewed in this perspective, a knowledge base may be equated to a collection of generalized assignment statements, and a query may be interpreted as a question regarding the value of a specified variable. Equivalently, a knowledge base may be regarded as a specification of elastic constraints on a collection of knowledge base variables $X_1, ..., X_n$; the answer to a query as the induced constraint on the variable in the query; and the inference process as the computation of the induced constraint on the query variable as a function of the given constraints on the knowledge base variables. In this view, the inference process resembles this solution of a nonlinear program [42,44].

In the following, our discussion of the problem of inference will be limited in scope. More specifically, we shall restrict our attention to disjunctive (i.e., possibilistic) assignment statements, since the inference rules for conjunctive statements can readily be derived by dualization, that is, replacing $\subset$ (is contained in) with $\supset$ (contains), and $\cap$ (intersection) with $\cup$ (union). Furthermore, we shall state only the principal rules of inference and will omit proofs.

In the rules stated below, $X, Y, Z, ...,$ are the constrained variables and $A, B, C, ...,$ are the constraining possibility distributions.

Entailment principle

$$\begin{array}{c} X \text{ is } A \\ \underline{A \subset B} \\ X \text{ is } B \end{array} \ . \tag{3.1}$$

Unary conjunctive rule

$$\frac{\begin{array}{c} X \ \textit{is} \ A \\ X \ \textit{is} \ B \end{array}}{X \ \textit{is} \ A \cap B} \ . \tag{3.2}$$

In the conclusion, $A \cap B$ denotes the intersection of A and B, which is defined by

$$\mu_{A \cap B}(u) = \mu_A(u) \wedge \mu_B(u) \,, \quad u \in U \ . \tag{3.3}$$

Binary conjunctive rule

$$\frac{\begin{array}{c} X \ \textit{is} \ A \\ Y \ \textit{is} \ B \end{array}}{(X,Y) \ \textit{is} \ A \times B} \ , \tag{3.4}$$

where $A \times B$ denotes the cartesian product of A and B, defined by

$$\mu_{A \times B}(u,v) = \mu_A(u) \wedge \mu_B(v) \ , \quad u \in \mathrm{U}, \quad v \in \mathrm{V} \ , \tag{3.5}$$

where U and V are the domains of X and Y, respectively.

Cylindrical extension rule

$$\frac{X \ \textit{is} \ A}{(X,Y) \ \textit{is} \ A \times V} \ , \tag{3.6}$$

where V is the domain of Y.

Projective rule

$$\frac{(X,Y) \ \textit{is} \ A}{X \ \textit{is} \ {}_X A} \ , \tag{3.7}$$

where ${}_X A$ denotes the projection A on the domain of X. The membership function of ${}_X A$ is defined by

$$\mu(u) = \vee_v \, (\mu_A(u,v)) \ , \tag{3.8}$$

where $\vee_v$ denotes the suprenum over $\mathrm{v} \in \mathrm{V}$.

Compositional rule

$$\frac{\begin{array}{c} X \ \textit{is} \ A \\ (X,Y) \ \textit{is} \ B \end{array}}{Y \ \textit{is} \ A \circ B} \ , \tag{3.9}$$

where $A \circ B$ denotes the composition of A and B, defined by

$$\mu_{A \circ B}(v) = \vee_u \, \mu_A(u) \wedge \mu_B(u,v) \ . \tag{3.10}$$

The compositional rule may be viewed as a corollary of the cylindrical extension rule, the binary conjunctive rule and the projective rule.

Extension principle

$$\frac{X \text{ is } A}{f(X) \text{ is } f(A)} , \tag{3.11}$$

where f is a function from U to V, and $f(A)$ is a possibility distribution defined by

$$\mu_{f(A)}(v) = \bigvee_u \mu_A(u) , \quad \text{over all } u \text{ such that } v = f(u). \tag{3.12}$$

A more general version of the extension principle which follows from (3.4) and (3.11) is

$$\begin{array}{l} X \text{ is } A \\ \underline{Y \text{ is } B} \\ f(X,Y) \text{ is } f(A,B) \end{array} . \tag{3.13}$$

Generalized modus ponens

$$\begin{array}{l} X \text{ is } A \\ \underline{\text{if } X \text{ is } B \text{ then } Y \text{ is } C} \\ Y \text{ is } A o (B' \oplus C) \end{array} , \tag{3.14}$$

in which B' is the complement of B and $\oplus$ is the bounded sum, defined by

$$\mu_{B' \oplus C}(v) = 1 \vee (1 - \mu_B(v) + \mu_C(v)) , \tag{3.15}$$

where $\vee$ = max. The inference rule expressed by (3.14) follows from the compositional rule of inference (3.9) and the assumption that the meaning of the conditional assignment statement which is the second premise in (3.14) is expressed by [36]

$$\text{if } X \text{ is } B \text{ then } Y \text{ is } C \rightarrow \pi_{(Y|X)}(u,v) = 1 \vee (1 - \mu_A(u) + \mu_B(v)) , \tag{3.16}$$

where $\pi_{(Y|X)}$ denotes the conditional possibility distribution function of Y given X.

REFERENCES AND RELATED PUBLICATIONS

1. Ballmer, T.T., and Pinkal, M. (eds.), *Approaching Vagueness*. Amsterdam: North-Holland, 1983.
2. Bandler, W., Representation and manipulation of knowledge in fuzzy expert systems, *Proc. Workshop on Fuzzy Sets and Knowledge-Based Systems*, Queen Mary College, University of London, 1983.
3. Bartsch, R. and Vennenmann, T., *Semantic Structures*. Frankfurt: Attenaum Verlag, 1972.
4. Barwise, J. and Cooper, R., Generalized quantifiers and natural language, *Linguistics and Philosophy 4* (1981) 159-219.
5. Bonissone, P.P., A survey of uncertainty representation in expert systems, in *Proc. Second Workshop of the North-American Fuzzy Information Processing Society*, General Electric Corporate Research and Development, Schenectady, NY, 1983.

6. Bosch, P., Vagueness, ambiguity and all the rest, in: *Sprachstruktur, Individuum und Gesselschaft*, Van de Velde, M., and Vandeweghe, W. (eds.). Tubingen: Niemeyer, 1978.
7. Brachman, R.J., What is-a is and isn't, *Computer 16* (1983).
8. Cresswell, M.J., *Logic and Languages*. London: Methuen, 1973.
9. Czogala, E., *Probabilistic Sets in Decision Making and Control*. Rhineland: Verlag TUV, 1984.
10. Dubois, D., and Prade, H., Fuzzy cardinality and the modeling of imprecise quantification, *Fuzzy Sets and Systems 16* (1985) 199-230.
11. Dubois, D., and Prade, H., *Théorie des Possibilités*, Paris: Masson, 1985.
12. Fox, M.S., On inheritance in knowledge representation, *Proc. IJCAI* (1979) 282-284.
13. Giles, R., Foundations for a theory of possibility, in: *Fuzzy Information and Decision Processes*, Gupta, M.M. and Sanchez, E. (eds.). Amsterdam: North-Holland, 183-195.
14. Goodman, I.R., and Nguyen, H.T., *Uncertainty Models for Knowledge-Based Systems*. Amsterdam: North-Holland, 1985.
15. Hirota, K., and Pedrycz, W., Analysis and synthesis of fuzzy systems by the use of probabilistic sets, *Fuzzy Sets and Systems 10* (1983) 1-13.
16. Israel, D., The role of logic in knowledge representation, *Computer 16* (1983) 37-41.
17. Kamp, H., A theory of truth and semantic representation, in *Formal Methods in the Study of Language*, Groenendijk, J.A. et al, (eds.), Mathematical Centre, Amsterdam, Tract 135, 1981.
18. Kandel, A., *Fuzzy Mathematical Techniques with Applications*. Reading: Addison-Wesley, 1986.
19. Kaufmann, A. and Gupta, M., *Introduction to Fuzzy Arithmetic*. New York: Van Nostrand, 1985.
20. Keenan, E., (ed.). *Formal Semantics of Natural Language*. Cambridge: Cambridge University Press, 1975.
21. Lambert, K., and van Fraassen, B.C., Meaning relations, possible objects and possible worlds, *Philosopical Problems in Logic* (1970) 1-19.
22. Mamdani, E.H., and Gaines, B.R., *Fuzzy Reasoning and its Applications*. London: Academic Press, 1981.
23. McDermott, D., and Cherniak, E., *Introduction to Artificial Intelligence*. Reading: Addison-Wesley, 1985.
24. Moore, R.C., *Problems in Logical Form*, SRI Tech. Report 241, Menlo Park, 1981.
25. Moore, R.C., The role of logic in knowledge representation and commonsense reasoning, *Proc. AAAI* (1982) 428-433.
26. Nilsson, N., *Principles of Artificial Intelligence*, Palo Alto: Tioga Press, 1980.
27. Ralescu, D., Toward a general theory of fuzzy variables, *Journ. Math. Analysis and Appl. 86* (1982) 176-193.
28. Rich, C., Knowledge representation languages and predicate calculus: how to save your cake and eat it too, *Proc. AAAI* (1982) 192-196.
29. Scheffler, I., *A Philosophical Inquiry into Ambiguity, Vaguenss and Metaphor in Language*. London: Routledge & Kegan Paul, 1981.
30. Tarski, A., *Logic, Semantics, Metamathematics*. Oxford: Clarendon Press, 1956.
31. van Fraassen, B.C., *Formal Semantics and Logic*. New York: Macmillan, 1971.
32. Wahlster, W., Hahn, W.V., Hoeppner, W., and Jameson, A., The anatomy of the natural language dialog system HAM-RPM, in: *Natural Language Computer Systems*, Bolc, L., (ed.). Amsterdam: North-Holland, 205-233, 1976.